SpringerBriefs in Plant Science

More information about this series at http://www.springer.com/series/10080

Shoib Ahmad Baba • Nasheeman Ashraf

Apocarotenoids of Crocus sativus L: From biosynthesis to pharmacology

 Springer

Shoib Ahmad Baba
Department of Plant biotechnology
Indian Institute of Integrative Medicine
Srinagar, Jammu and Kashmir, India

Nasheeman Ashraf
Department of Plant biotechnology
Indian Institute of Integrative Medicine
Srinagar, Jammu and Kashmir, India

ISSN 2192-1229   ISSN 2192-1210   (electronic)
SpringerBriefs in Plant Science
ISBN 978-981-10-1898-5   ISBN 978-981-10-1899-2   (eBook)
DOI 10.1007/978-981-10-1899-2

Library of Congress Control Number: 2016948214

Printed on acid-free paper

This Springer imprint is published by Springer Nature
The registered company is Springer Nature Singapore Pte Ltd.
The registered company address is: 152 Beach Road, #22-06/08 Gateway East, Singapore 189721, Singapore

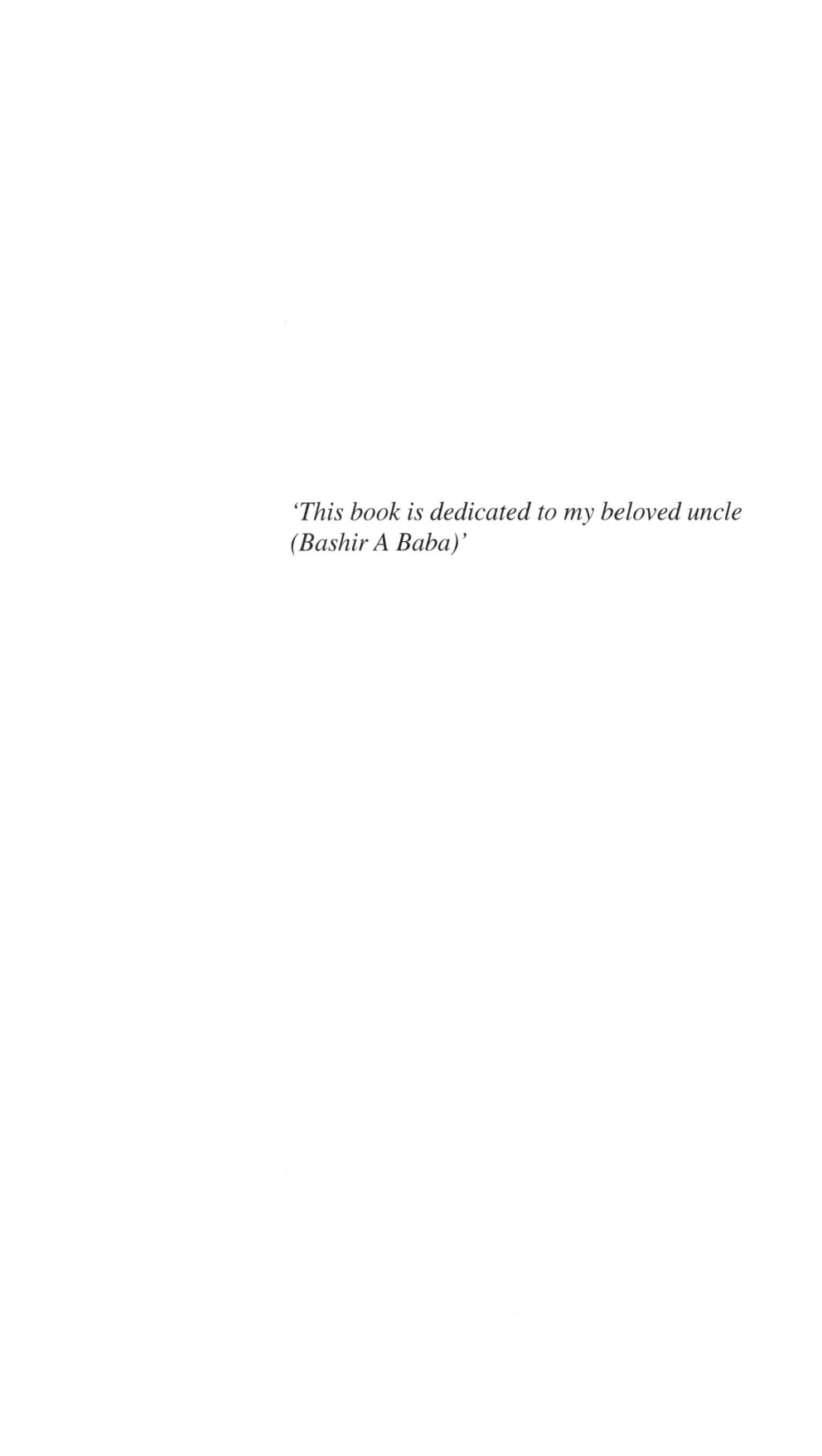

'This book is dedicated to my beloved uncle (Bashir A Baba)'

Contents

About the Authors

Dr. Nasheeman Ashraf is currently a scientist at the Indian Institute of Integrative Medicine's Department of Plant Biotechnology. She completed her PhD at the National Institute of Plant Genomic Research (NIPGR), New Delhi in 2010, and then joined the Indian Institute of Integrative Medicine, Srinagar (CSIR-IIIM) in 2012. Her research has been published in respected journals, including BMC Genomic, BMC Plant biology, PLOS One, the Journal of Plant physiology. She is a member of the American Society of Plant Biologists. Her current research interest is *Crocus sativus*, and she was involved developing the transcriptome map for *Crocus* in 2014. This book has been based on years of research carried out in Dr. Nasheeman's lab.

Shoib Ahmad Baba completed his master's in botany at the University of Kashmir in 2011, and joined the Indian Institute of Integrative Medicine, Srinagar in 2012 to pursue research under the supervision of Dr. Nasheeman Ashraf. He has been working on *Crocus sativus* from last 5 years and has identified and characterized some of the key genes of apocarotenoid and flavonoids biosynthesis. Shoib Ahmad Baba's research has been published in leading journals, such as BMC Genomics and the Journal of Plant Physiology. He was actively involved in developing the transcriptome map of *C. sativus* in Dr. Nasheeman's lab. Shoib Ahmad Baba is currently a member of the 'Insights in Genes and Genomes' editorial board.

Chapter 1
Apocarotenoid Biosynthesis in *Crocus sativus* L.

Abstract Apocarotenoids are oxidative cleavage products of carotenoids. These molecules play vital physiological and developmental roles in plants. Besides this, apocarotenoids also hold tremendous pharmacological importance. Apocarotenoids are ubiquitously found across plant kingdom, but *Crocus sativus* (saffron) is the only source of some unique and economically important apocarotenoids. These apocarotenoids include crocin, picrocrocin, and safranal which besides having pharmacological importance are also responsible for the color, flavor, and aroma of the world's costliest spice (saffron). Apocarotenoid biosynthesis in *C. sativus* is regulated throughout the life cycle with active changes in apocarotenoid composition of stigma due to developmental stage-specific requirements and in response to external environmental cues. Although the biosynthesis of these unique *C. sativus* apocarotenoids has been elucidated to a greater extent, there are still some missing links in the pathway. Besides, only a few studies have been carried out on the regulation, tissue- and developmental-specific accumulation, and transport of apocarotenoids in *C. sativus* as well as in other plants. The present review is an organized attempt to gain insights about the biosynthesis, regulation, and transport of apocarotenoids in *C. sativus*.

Keywords *Crocus sativus* • Apocarotenoids • Crocin • Safranal • Carotenoid cleavage dioxygenase

1.1 Introduction

Carotenoids are the most prevalent group of pigments found in nature (Britton et al. 2009; Rao and Rao 2007). These molecules contribute to diversity of colors found in many flowers, fruits, and vegetables which is attributed to the presence of a number of conjugated double bonds in a polyene chain functioning as a chromophore. Carotenoids are synthesized not only in all photosynthetic organisms but also in some non-photosynthetic bacteria and fungi. Animals can't synthesize carotenoids; however, through dietary intake of carotenoid-containing foods, they accumulate carotenoids in their bodies that contribute to their health and behavior. For instance, some fish and birds accumulate dietary carotenoid which boosts their immune

© Springer Science+Business Media Singapore 2016   1
S.A. Baba, N. Ashraf, *Apocarotenoids of Crocus sativus L: From biosynthesis to pharmacology*, SpringerBriefs in Plant Science, DOI 10.1007/978-981-10-1899-2_1

system (Cazzonelli and Pogson et al. 2010). The human health benefits associated with carotenoids are well reported in literature (DellaPenna and Pogson 2006; Krinsky and Johnson 2005; Davies 2007). In brief, carotenoids promote antioxidant activity, act as precursors for vitamin A, and decrease the risk of diseases like cancer and neurodegeneration (Halliwell 2006, 2007; Cazzonelli and Pogson et al. 2010). In plants, carotenoids play several vital functional roles in development, photosynthesis, and membrane stability (Cazzonelli and Pogson et al. 2010; Baba et al. 2015a). They also act as precursors for the biosynthesis of phytohormones and also have some ecological roles such as attracting pollinators and seed dispersers (Schwartz et al. 2004). Carotenoids are not only important in their intact form; they also undergo cleavage to form an equally important group of compounds called apocarotenoids. The cleavage may be as a result of oxidative damage or an enzymatic reaction catalyzed by a group of enzymes called carotenoid cleavage dioxygenases (CCDs). CCDs carry out the cleavage of carotenoids through a reaction involving incorporation of both oxygen atoms from molecular oxygen across a double bond resulting in the production of two ketone or aldehyde-containing cleavage products (Auldridge et al. 2006a, b).

Crocus sativus, a triploid geophyte, is an important source of unique apocarotenoids including crocin, picrocrocin, and safranal which are responsible for color, flavor, and aroma of spice, respectively (Ashraf et al. 2015). Besides *C. sativus*, these apocarotenoids have been detected only in a few plants species such as *Buddleja* (Liao et al. 1999) and *Gardenia* (Pfister et al. 1996). The biosynthesis of *C. sativus* apocarotenoids occurs through methylerythritol 4-phosphate (MEP) pathway and starts with the cleavage of zeaxanthin, to produce cyclic carotenoid VOCs (picrocrocin and safranal) and crocetin, which is eventually glycosylated to crocin (Fig. 1.1). Apocarotenoid biosynthesis in *C. sativus* is a tissue and developmental stage-specific process. The pathway involved is highly regulated and is coordinated effort of many individual pathways. Although a number of genes involved apocarotenoid biosynthesis in *C. sativus* were identified previously, there were still some missing links in the whole apocarotenoid biosynthesis. It was not until recently that Baba et al. (2015b) reconstructed the whole apocarotenoid biosynthesis pathway by comprehensive transcriptome sequencing (Baba et al. 2015b). Further, only a little is known about the regulation of apocarotenoid biosynthesis and transport of apocarotenoids in *C. sativus*. The present review will therefore mainly focus on apocarotenoid biosynthesis with emphasis on how carotenoid pathway supplies substrates for biosynthesis of apocarotenoids and how the latter are modified by enzymes to carry out different functions. Further, this review will also attempt to summarize the recent advancements made in understanding the regulation of apocarotenoid biosynthesis and their subsequent transport in *C. sativus*

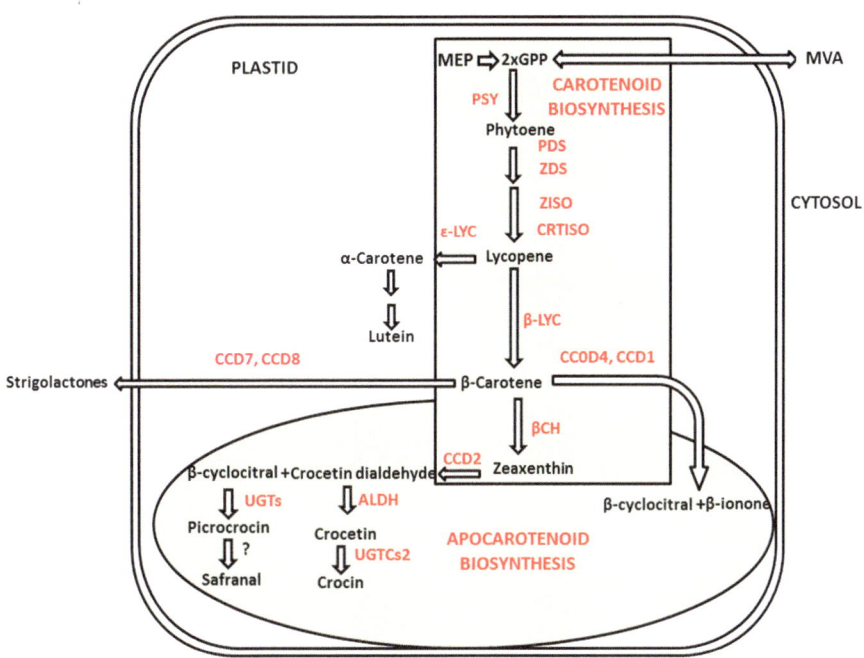

Fig. 1.1 Apocarotenoid biosynthesis pathway in *C. sativus* L.

1.2 Apocarotenoid Biosynthesis in *C. sativus* L.

1.2.1 *Apocarotenoid Biosynthesis Depends on the Availability of Carotenoid Substrates*

Carotenoids are prevalent molecules of ancient origin with a diversity of functions. These molecules emerged early on in the microbial world, including Archaea, with functions of membrane stabilizers and UV protectants. Almost all carotenoids stem from a linear tetraterpene polyene (C_{40}), wherein the presence of conjugated double bonds confers the light-absorbing property. Carotenoids have developed a number of functions in higher organisms due to their ubiquitous occurrence and therefore their accessibility to develop new functions. Examples include the evolution of several signaling molecules from carotenoid precursors and their oxygenated derivatives. Oxidative tailoring of carotenoids to specific regulatory cleavage products (apocarotenoids) catalyzed by a multienzyme family of structurally conserved carotenoid cleavage dioxygenases (CCDs) is a recurrent theme throughout evolution from bacteria to mammals (Walter and Strack 2011; Sauter et al. 2001; Farré-Armengol et al. 2013). Apocarotenoid biosynthesis is an extension of carotenoid biosynthetic pathway and is limited by the rate at which carotenoid pathway supplies the substrates. Zeaxanthin and β-carotene are the two important carotenoids which act as substrates for the biosynthesis of unique apocarotenoids of *C. sativus* (Rubio et al. 2008; Castillo et al. 2005). Carotenoids are synthesized de novo from the isomeric C_5 precursors, isopentenyl diphosphate (IDP) and dimethylallyl diphosphate (DMAPP) in plastids. Both of these C_5 precursors originate primarily from the plastidial methylerythritol 4-phosphate (MEP) pathway. The condensation of three IPP units and one molecule of DMAPP leads to the production of geranyl-geranyl diphosphate (GGPP, C_{20}). This reaction is catalyzed by the enzyme GGPP synthase (GGPPS). Zeaxanthin biosynthesis in *C. sativus* starts with geranylgeranyl diphosphate, followed by the condensation of two molecules of geranylgeranyl diphosphate by phytoene synthase (PSY) which gives rise to 15-*cis*-phytoene, the first specific compound in carotenoid biosynthesis. This is converted into bright red all-*trans* lycopene which is the first pigmented carotenoid in a series of four desatu-ration reactions. These desaturation reactions are catalyzed by phytoene desaturase (PDS), ζ-carotene isomerase (Z-ISO), ζ-carotene desaturase (ZCD), and carotenoid isomerase (CRTISO). Phytoene is converted to lycopene PDS and ZCD (Fig. 1.1). This pathway gives rise to poly-*cis* compounds which are converted to their all-*trans* forms by CRTISO (Cazzonelli and Pogson et al. 2010). CRTISO has emerged as a regulatory node in the carotenoid biosynthesis pathway. CRTISO mutants, such as *ccr2* and *tangerine*, result in accumulation of *cis*-carotenes which include 7-7′,9-9′-tetra-*cis*-lycopene, in the seedling etioplasts and tomato fruit chromo-plasts. Notwithstanding this block in etioplasts and chromoplasts, the biosynthetic pathway continues in chloroplasts through photoisomerization. However, there is delayed greening and significant reduction in lutein in *Arabidopsis* and variable degrees of chlorosis in tomato and rice (Isaacson et al. 2002, 2004; Li et al. 2007, 2008; Park et al. 2002; Dong et al. 2007; Yu et al. 2007; Corona et al. 1996; Cazzonelli et al. 2009; Fang et al. 2008). It should be noted that CRTISO was identi-fied recently isolated from *C. sativus* and requires further characterization (Baba et al. 2015b). Lycopene acts as the substrate for two competing enzymes, lycopene β-cyclase (β-LCY) and lycopene ε-cyclase (ε-LCY), and represents a branch point of the carotenoid pathway. Both β-LCY and ε-LCY cyclize the linear backbone to create terminal ionone rings. However, the structures of these rings are distinct from each other. The addition of one β-ring to lycopene by β-LCY produces ɣ-carotene, and the addition of a second β-ring to the free end by the same enzyme produces the orange pigment β-carotene. Two β-ring hydroxylations of β-carotene yield zeaxan-thin in a reaction catalyzed by β-carotene hydroxylase (BCH) (Castillo et al. 2005). Both β-carotene and zeaxanthin act as precursors for the biosynthesis of unique apocarotenoids of *C. sativus*.

1.2.2   Carotenoid Cleavage Leads to the Production of Apocarotenoids

The repeated structural motifs found in the large diversity of well-known apocarot-enoids point out that they are derived from full-length carotenoids, probably along common biosynthetic pathways (Kloer and Schulz 2006) (Fig. 1.2). In *C. sativus*,

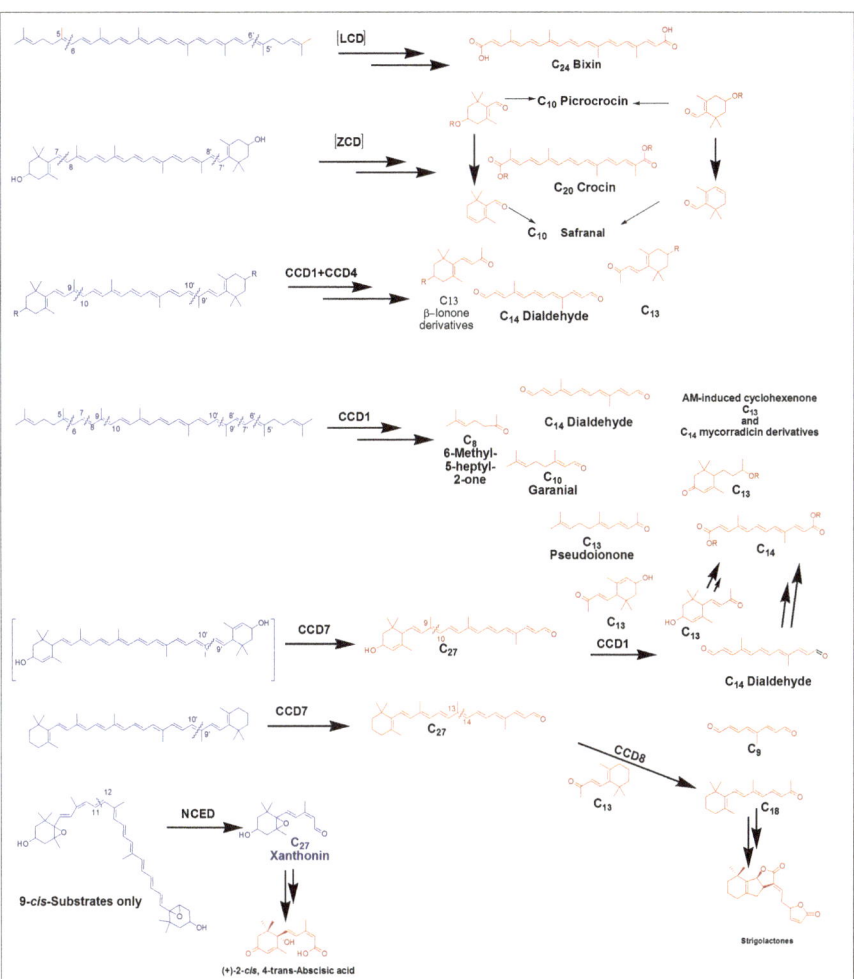

Fig. 1.2 Different CCDs recognize different cleavage sites leading to the production of diversity of apocarotenoids

apocarotenoids are formed by the cleavage of carotenoids (zeaxanthin and β-carotene mainly) by carotenoid cleavage dioxygenases (CCDs) (Rubio et al. 2008). This family is generally characterized by an amphipathic α-helix domain, a catalytic domain which contains four conserved histidine residues responsible for nonheme iron coordination, Fe^{2+} as a cofactor, and a plastid-targeting transit peptide except for some CCDs that are localized in cytoplasm like CCD1 (Kloer and Schulz 2006; Messing et al. 2010). CCDs of plant and animal origin are presumed to act similarly in that they incorporate both oxygen atoms from molecular oxygen into their substrates across a double bond resulting in the production of two aldehyde-containing cleavage products. There has been a strong debate regarding the monooxygenase and the dioxygenase mechanism of these enzymes; however, there is a convincing evidence now for a dioxygenase mechanism obtained from labeling experiments (Auldridge et al. 2006a; Sui et al. 2013). Recently, the crystal structure of apocarotenoid-15,15′-oxygenase (ACO) from cyanobacterium *Synechocystis* was successfully determined, and this enzyme was found to have seven-bladed beta-propeller tertiary structure. The ACO protein crystal structure has shown that the enzyme holds a Fe^{2+} ion coordinated to four conserved histidine residues in the active site. This four-histidine–iron arrangement is shared by all members of CCD multienzyme family. However, except from this characteristic, their amino acid sequences are somewhat heterogeneous. It has also been reported that during catalysis, there appears a change of three consecutive double bonds from all *trans* to a cranked *cis–trans–cis* configuration, in which the remaining *trans*-bond is located at the dioxygen-ligated Fe^{2+} and cleaved by oxygen (Kloer and Schulz 2006; Poliakov et al. 2005). The plant CCDs show high degree of similarity with bacterial lignostilbene dioxygenases (LSD) in overall sequence, conservation of histidine residues, iron as a cofactor, and cleavage of double bond suggesting that the enzymes might have evolved from a common ancestor (Kamoda and Saburi 1993; Bugg et al. 2011; Tan et al. 1997). Recent studies on some CCDs have shown the presence of intrinsic isomerase activity including RPE65 and NinaB and mammalian CCDs in general (Moiseyev et al. 2006; Oberhauser et al. 2008). However, no intrinsic isomerase activity has been seen in plant CCDs which could be due to the independent presence of iron-dependent carotenoid isomerases (Auldridge et al. 2006a). The first carotenoid cleavage dioxygenases involved in apocarotenoid biosynthesis that were cloned, identified, and functionally characterized in *C. sativus* include zeaxanthin cleavage dioxygenase (*Cs*ZCD) and carotenoid (9′–10′) cleavage dioxygenase (*Cs*CCD) (Bouvier et al. 2003, 2005). The expression of *Cs*ZCD was found to be restricted to the style branch tissues, and it is enhanced under dehydration stress, whereas *Cs*CCD is expressed in a constitutive manner in flower and leaf tissues irrespective of dehydration stress. Studies have revealed the presence of a stepwise sequence including the oxidative cleavage of zeaxanthin inside the chromoplasts followed by the sequestration of water-soluble cleavage products into the central vacuole (Bouvier et al. 2003). Recently, two isoforms of *Cs*CCD1 and three isoforms of *Cs*CCD4 were also identified and functionally characterized in *C. sativus* and found to be involved in the biosynthesis of β-ionone. It was observed that *Cs*CCD1a exhibited constitutive expression; *Cs*CCD1b expression was restricted to the stigma only. However, *Cs*CCD4a, *Cs*CC4b, and *Cs*CCD4c had expression patterns in accordance with the emission of β-ionone derived from β-carotene during the stigma development. The function of *Cs*CCD4 isoforms in the production β-ionone was confirmed by overexpressing them in bacterial strains accumulating β-carotene (Rubio et al. 2008). Besides *C. sativus*, CCDs also control the emission of β- and α-ionone in other plants like *Nypa fruticans* and *Lycopersicon esculentum* (Simkin et al. 2004a, b; Azuma et al. 2002). The apocarotenoid β-ionone has been reported to carry out some essential functions in plants. While as β-ionone specifically repels beetles (*Anomala transvaalensis*), α-ionone is more active against *Macrodactylus subspinosus*, which can cause serious damage to the plants

(Donaldson et al. 1990; Flath et al. 1994 McQuate and Peck 2001). *Halotydeus destructor* fed less on *Trifolium glanduliferum* extracts that had high levels of β-ionones and other terpenes (Wang et al. 2004). A groundbreaking study, conducted by Wei et al. (2011), showed that transgenic *Arabidopsis* overexpressing CCD1 exhibited enhanced levels of α-ionone, a volatile apocarotenoid produced as a result of CCD1 dioxygenase activity. Moreover, when transgenic plants with elevated levels of α-ionone were tested for their interaction with *Phyllotreta cruciferae* (crucifer flea beetles), feeding damage was reduced in the transgenic plants compared to wild-type plants (Wei et al. 2011). This suggested that the volatile apocarotenoids deterred the insects from feeding on these plants. These compounds also shield the male gametophyte from pollen-consuming organisms that do not provide any benefit to plant fitness (Farré-Armengol et al. 2013).

Recently, a new carotenoid cleavage dioxygenase CCD2 was isolated from *C. sativus* stigma (Frusciante et al. 2014). Although *Cs*CCD2 has different activities than *Cs*CCD1 and *Cs*CCD4 isoforms, phylogenetic analysis revealed it to be closely related to *Cs*CCD1.

*Cs*CCD7 and *Cs*CCD8 have also been identified from *Crocus sativus*. The sequence and phylogenetic analysis of *Cs*CCD7 and *Cs*CCD8 result suggests that CCD7/CCD8 genes had similar evolutionary trends than CCD1 and CCD4 subfamily. Further, phylogenetic analysis of *Cs*CCD7 and *Cs*CCD8 also provides a clue about CCD duplication, which ultimately led to the emergence of two lineages that evolved into CCD7 and CCD8. In *C. sativus*, *Cs*CCD8 is highly expressed in quiescent axillary buds, and its expression is significantly reduced by decapitation suggesting its involvement in the suppression of axillary bud outgrowth. Unlike *Cs*CCD8, *Cs*CCD7 is highly expressed in the newly developed vascular tissue of axillary buds compared to the vascular tissue of the apical bud (Rubio-Moraga et al. 2014). While as the substrates of CCD7 and CCD8 have not been determined in *C. sativus*, in other plants, they have been reported to cleave β-carotene sequentially leading to the production of strigolactone, a novel apocarotenoid hormone involved in shoot branching (Umehara et al. 2008; Gomez-Roldan et al. 2009). Strigolactones, first discovered in root exudates of root-parasitic plant *Striga hermonthica*, produced under the nutrient-limiting conditions in roots, promote the growth of lateral roots and root hairs and inhibit the growth of lateral branches and buds in shoot (Umehara et al. 2008). This strategy increases the nutrient uptake by the roots while at the same time reduces the shoot's demand for the resources. Strigolactones have been reported to have an effect, exclusively or in blend with other phytohormones, in the morphology of potato plants and also in directing stolon development and maintaining tuber dormancy. This is supported by the studies carried on CCD8-RNAi potato plants which had more lateral and main branches than control plants and reduced stolon formation, besides exhibiting a dwarfing phenotype and a lack of flowering in the most severely affected lines (Pasare et al. 2013).

1.2.3   Apocarotenoids in C. sativus Are Stored in Glycosylated Form

Plants synthesize a wide array of secondary metabolites with a diversity of functions. Many of these secondary metabolites are stored in a glycosylated form which chemically stabilizes and renders them suitable for storage in different organelles. In *C. sativus*, *Cs*CCD2 leads to the production of a crocetin and hydroxy-β-cyclocitral. Both of these are glucosylated by glycosyltransferases to form crocin and picrocrocin, respectively. Crocin, the most important apocarotenoid of *C. sativus* has different forms depending on the number of glycosylations. The crocin family consists of glucosyl and gentiobiosyl esters of crocetin, a dicarboxylic 20-carbon carotenoid (Verma and Middha 2010). Crocin quickly dissolves in water to form orange-colored solution, and this is the reason why saffron is used as a natural food colorant. As discussed above, depending on the number of glycosylations, different types of crocins have been identified (Fig. 1.3). Verma and Middha (2010) identified five different forms of crocin designated as crocins 1, 2, 3, 4, and 5 (Verma and Middha 2010). Similarly, two types of crocins were isolated by Pfister et al. (1996). These included crocetin di(β-gentiobiosyl) ester and crocetin mono(β-gentiobiosyl) ester (Pfister et al. 1996). Crocin analogues including crocins 1–4 are mostly glycosides of *trans*-crocetin in saffron with *trans*-crocins 3 and 4 as the most abundant ranging between 0.01–9.44 % and 0.46–12.12 %, respectively (Alonso et al. 2001; Pfander and Wittwer 1975). Unlike *trans*-crocins, *cis*-crocins are present only in minor amounts (Li et al. 1999). Except for crocin 1, all the other forms have been reported to be present in both *cis*- and *trans*-isoforms (Dhingra et al. 1975). It has been shown that *trans*-crocins undergo photoisomerization reactions and convert to *cis*-crocins; however, this process is dependent on the agricultural and environmental conditions in which the plant grows (Speranza et al. 1984). Tarantilis et al. (1995) reported that compared to the other carotenoids of saffron, all-*trans* crocetin

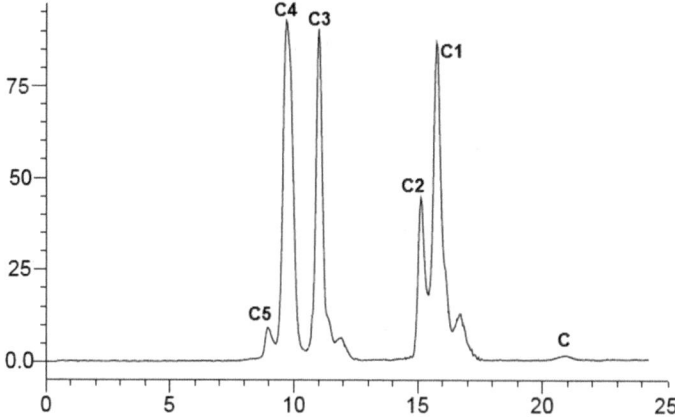

Fig. 1.3 HPLC showing five different types of peaks representing five different crocins

di(β-D-gentiobiosyl) ester shows the maximum coloring capacity which is due to its high water solubility (Tarantilis et al. 1995). Cote et al. (2001) reported that the glucosylation of crocetin into crocin involves two glucosyltransferases: GTase1 which catalyzes the glucose transfer on both carboxylic ends of crocetin and GTase2 which helps glucosidic bonds to form gentiobiosyl esters. A partially purified UDP-glucose–crocetin glucosyltransferase was found to be involved in crocin biosynthesis by forming ester bonds between the carboxylic groups of crocetin and the glucose moiety of UDP-glucose (Cote et al. 2001). Moreover, a GTase1 activity was also reported later in the studies carried out by Yang et al. (2005). They reported GTase1 activity to be higher during the first 4 days of crocin glucoside biosynthesis; however, significant decrease was reported after the fourth day. A full-length crocetin glucosyltransferase UGT*Cs*2 gene was first time isolated, expressed in bacteria, and functionally characterized by Moraga et al. (2004). The recombinant protein UGT*Cs*2 had glucosylation activity against crocetin, crocetin β-D-glucosyl ester, and crocetin β-D-gentiobiosyl ester. Besides crocetin, hydroxy-β-cyclocitral that is formed concomitantly with crocetin is glucosylated to picrocrocin, the bitter flavoring component of saffron. Picrocrocin accounts for approximately 1–13 % of saffron's dry matter (Alonso et al. 2001). However, till date glucosyltransferase involved in the glucosylation of β-cyclocitral has not been identified. The isolation of these enzymes is of importance to food industry. Their biochemical characterization may help to make decisions about whether to add them to drinks and beverages before or during processing in order to enhance their flavor, color, or other quality factors.

1.2.4   Hydrolysis of Picrocrocin to Safranal

For the first time, safranal was isolated and identified from saffron in the twentieth century as the major aromatic compound, and its structure was first time determined in 1932 (Khun and Winterstein 1934). Although safranal (30–70 % of essential oil and 0.001–0.006 % of dry matter) is the major aromatic apocarotenoid in *C. sativus*, other aromatic compounds, like isophorone, 2,2,6-trimethyl-1,4-cyclohexanedione, 4-ketoisophorone, 2-hydroxy-4,4,6-trimethyl-2,5-cyclohexadien-1-one, as well as 2,6,6-trimethyl-1,4-cyclohexadiene-1-carboxaldehyde, also add to the aroma of the saffron (Maggi et al. 2009). Till date more than 160 additional volatile components have been identified from *C. sativus*, and many more wait identification (Carmona et al. 2006). Safranal is a hydrolysis product of picrocrocin. Although safranal has been obtained by both acid/alkaline hydrolysis (Winterstein and Teleczky 1922) and enzymatic hydrolysis (Khun and Winterstein 1934) of picrocrocin under in vitro conditions, whether its mechanism is operative under in vivo conditions still remains doubtful. However, studies carried from time to time have considered dehydration important for the production of safranal from picrocrocin. The first study to point out the important role of dehydration in the generation of aromatic components of saffron including safranal was carried out by Curro et al. 1986. They observed a great qualitative and quantitative difference in dehydrated and non-dehydrated saffron. The lower number of compounds present in non-dehydrated saffron was believed to be due to greater water content (Curro et al. 1986). Similar results were obtained by Himeno and Sano (1987) when investigating the safranal and 4-β-hydroxysafranal content in stigma-like structures obtained in vitro starting from young *C. sativus* ovaries. They proposed that the formation of safranal starts from 4-β-hydroxysafranal or even from picrocrocin. This transformation would be mainly promoted by the dehydration process (Himeno and Sano 1987). Raina et al. (1996) reported that warm or vacuum drying increased the safranal content of saffron (Raina et al. 1996). Gregory et al. (2005) reported 70–90 °C as the best temperature for highest safranal generation during drying (Gregory et al. 2005). It is therefore presumed that precursors such as 4-β-hydroxysafranal are formed during the dehydration processes, which then give rise to safranal. Since, safranal is an essential component of saffron aroma, the mechanism by which it is produced from picrocrocin needs to be fully elucidated. Whether it is produced nonenzymatically, enzymatically, or by both the processes needs to be understood.

1.3   Regulation of Apocarotenoid Biosynthesis in *C. sativus*

Identification of mechanisms and processes by which carotenoid biosynthesis and its degradation is regulated has been one of the greatest challenges in the studies of carotenoid metabolism. During the course of evolution, plants have developed complex regulatory mechanisms controlling biosynthesis and accumulation of these molecules. The essential roles of carotenoids and their cleavage derivatives in various facets of plant physiology and development suggest that their biosynthesis is coordinately regulated with other processes such as plastid biogenesis, flowering, and fruit development (Fraser and Bramley 2004). In *C. sativus*, apocarotenoid biosynthesis is regulated by the expression of key carotenoid biosynthesis pathway genes and carotenoid cleavage dioxygenases. Besides these, some transcription factors have also been identified that regulated the expression of genes involved in both carotenoid and apocarotenoid biosynthesis.

1.3.1   *Regulation of Apocarotenoid Biosynthesis by Carotenogenic Enzyme Levels*

In plants transcriptional regulation of carotenoid gene expression appears to be a major mechanism by which the biosynthesis and accumulation of specific carotenoids or their derivatives are regulated during flower color development and fruit ripening. The biogenesis of the three major carotenoid derivatives, crocin, picrocrocin, and safranal, occurs by the oxidative cleavage of zeaxanthin and β-carotene.

The accumulation of these carotenoids in *C. sativus* appears to be controlled by the transcript levels of PSY, BCH, and β-LCY2a. Castillo et al. (2005) reported that with the transition of yellow undeveloped to red fully developed stigmas of *C. sativus*, accumulation of zeaxanthin was observed to be associated with increased expression of PSY (Castillo et al. 2005). PSY is generally accepted as the key regulatory enzyme in the carotenoid biosynthesis pathway. It has been reported that PSY genes respond to high light, salt, photoperiod, temperature, drought, development signals, and posttranscriptional feedback regulation. Although in *C. sativus* only one PSY gene has been identified till now, there are two or more homologues in tomato, rice (*Oryza sativa*), poplar (*Populus trichocarpa*), maize (*Zea mays*), bread, and wheat (*Triticum aestivum*). The activity of these multiple PSY enzymes appears redundant, but their expression is tissue specific (Welsch et al. 2008; Chaudhary et al. 2010; Howitt et al. 2009).

Carotenoid biosynthesis splits after lycopene to form epsilon- and beta-carotenoids by activity of two lycopene cyclases, Ɛ-LCY and β-LCY. This branch point is reported to have a main regulatory role in controlling the ratio of the most abundant carotenoid, lutein, to the beta-carotenoids (Cazzonelli and Pogson et al. 2010). In *C. sativus*, two different lycopene cyclase genes have been identified (β-LCY1 and β-LCY2a), and the expression of β-LCY2a in different *Crocus* species indicates that transcriptional regulation of this gene has profound effects on the quantitative accumulation of apocarotenoids. A recent study has revealed a sharp increase in the expression of β-LCY2a during the transition of yellow to scarlet stage of stigma which is associated with increase in the total carotenoid content.

Besides PSY and β-LCY2a, BCH has been shown to exhibit a regulatory role on the biosynthesis of *Crocus* apocarotenoids. It has been reported that accumulation of BCH transcripts at high levels in saffron stigma is associated with significant increase in the production and accumulation of the zeaxanthin-derived apocarotenoids crocin, crocetin, and picrocrocin (Bouvier et al. 2003; Castillo et al. 2005; Ahraze et al. 2009). In another study carried on stigma generated from tissue culture, BCH was found to be the key enzyme in apocarotenoid biosynthesis under in vitro conditions (Namin et al. 2009). Taken together, these studies provide strong evidence that reactions catalyzed by PSY, BCH, and β-LCY2a may be limiting steps for precursor availability for apocarotenoid biosynthesis in *C. sativus* and may be considered as important candidates for manipulation of apocarotenoid biosynthesis in *C. sativus*.

1.3.2   Regulation of Carotenoid Degradation

Carotenoid cleavage dioxygenase play an important role in carotenoid turnover. The expression of CCD enzymes determine the carotenoid as well as apocarotenoid content of many important crop plants, for instance, carotenoid content in *Solanum tuberosum* is completely dependent on the expression of carotenoid cleavage dioxygenase 4. Similarly, CCD4 is the major determinant in the accumulation of carotenoids and carotenoid-derived volatiles in peach fruit flesh (Brandi et al. 2011). Recent studies have shown that inactivation of CCD4 leads to the enhanced accumulation of carotenoids in *Prunus persica* (Ma et al. 2014).

Metabolic turnover of carotenoids by CCDs not only helps to maintain the steady level of carotenoids in plants but also leads to the production of important signaling and accessory apocarotenoid molecules. Further, the biogenesis of apocarotenoids not only depends on the availability of carotenoid substrates but also on the expression of CCDs. It has been observed that the expression CCDs directly coincides with the accumulation of apocarotenoids. In *C. sativus*, *Cs*CCD2 is expressed exclusively in the stigma, and therefore, biosynthesis and accumulation of crocetin and its glucosides are limited to the stigma only. Similarly, *C. sativus* CCD4 isoforms are expressed in the floral tissues and so is the production of β-ionone (Frusciante et al. 2014; Rubio et al. 2008). Moreover, there are three isoforms of *Cs*CCD4 in *C. sativus* which are all involved in the production of β-ionone, but each of these CsCCD4 isoforms shows a different expression pattern. While *Cs*CCD4a is expressed constitutively, *Cs*CCD4b and CsCCD4c are expressed in stigma only. This ensures the production of apocarotenoids at particular development stage and tissue. There are a number of convincing evidences that oxidative cleavage of carotenoids is induced under environmental stress. In *Petunia hybrida*, PhCCD1 involved in the production of β-ionone is induced by light and exhibits a circadian rhythm in both leaves and flowers (Simkin et al. 2004b). We recently reported that the three isoforms of *Cs*CCD4, although structurally similar, express under different stress conditions. The three isoforms showed no response to hormonal treatments, but expressed differentially under stress treatments like ultraviolet (UV) irradiation, methyl viologen (MV), cold (4 °C), and mannitol, suggesting that they come into play under different stresses or environmental clues (Baba et al. 2015c).

1.3.3   Transcription Factors Regulating Apocarotenoid Biosynthesis

Coordinated transcriptional regulation of biosynthetic genes is one of the major mechanisms responsible for the final levels of secondary metabolites in plant cells. This regulation of biosynthetic pathways is achieved by specific transcription factors (Vom Endt et al. 2002). Transcription factors are sequence-specific DNA-binding proteins that control the coordinated expression of genes by interacting with the promoter regions of target genes. Although, regulation flavonoid and terpenoid biosynthesis by transcription factors has been studied extensively, knowledge about the regulation of carotenoid and apocarotenoid expression by transcriptions factors is still rudimentary. Only a few transcription factors regulating apocarotenoid biosynthesis have been identified and characterized in plants. These studies include regulation of PSY gene expression by phytochrome-interacting factor

(PIF1) which regulates carotenoid accumulation during daily cycles of light and dark in mature plants by directly binding to the promoter of PSY (Toledo-Ortiz et al. 2010). In other studies, AP2 gene family (RAP2.2) has been reported to bind to PSY promoter and modestly regulate the transcript levels of PSY and PDS in *Arabidopsis* (Welsch et al. 2007). In *C. sativus*, the founding experiments to understand the regulation of apocarotenoids by transcription factors were recently carried out by our group. A transcription factor Ultrapetala 1 (*Cs*ULT1) was identified from the stigma of *C. sativus*. It was observed that *Cs*ULT1 transcript levels were more abundant in stigma and the expression increased from pre-anthesis stage till anthesis and decreased in post-anthesis stage. This expression of *Cs*ULT1 corroborated with the accumulation pattern of crocin indicating its possible role in regulation of apocarotenoid biosynthesis. To further confirm the role of *Cs*ULT1, it was overexpressed in *Crocus* calli which resulted in enhanced expression of key pathway genes like phytoene synthase (PSY), phytoene desaturase (PDS), beta-carotene hydroxylase (BCH), and carotenoid cleavage dioxygenases (CCDs). It is strongly believed that these experiments will form the basis for identification and characterization of transcription factors regulating apocarotenoid biosynthesis. Further, with the identification of transcription factor families such as MYB and MYC from *C. sativus* (Baba et al. 2015b), it is imperative that more research efforts in the future will go into their characterization and evaluation of their role in regulation of apocarotenoid biosynthesis.

1.4   Site of Synthesis and Transport of Apocarotenoids in *C. sativus*

In *C. sativus*, the unique apocarotenoids are synthesized mainly in the stigma of the flower. The main apocarotenoids that accumulate throughout stigma development in *C. sativus* are crocetin, its glucoside derivatives, crocins, and picrocrocin. The accumulation of these molecules begins early in the development of the stigma. Crocetin is detected in the earlier stages, and concentrations rapidly increase to a maximum in the red-immature stage of stigma development. Thereafter, crocetin content starts decreasing till anthesis. The crocins follow the same pattern as crocetin, but their content remains high in the mature stages until the time of anthesis (Fig. 1.4) (Rubio et al. 2008; Ashraf et al. 2015).

In order to understand the stigma-specific accumulation of apocarotenoids in *C. sativus*, expression profiling of main carotenoid and apocarotenoid biosynthesis genes was carried out in *C. sativus* stigma and flower minus stigma. It was found that expression of main carotenoid and apocarotenoid biosynthesis genes was significantly higher in stigma as compared to the rest of the flower which provided the molecular basis for the stigma-specific accumulation of *C. sativus* apocarotenoids. The expression of the genes such as PSY and PDS which catalyze initial reactions in carotenoid biosynthetic pathway is higher in stigma. Moreover, Z-ISO and ZCD

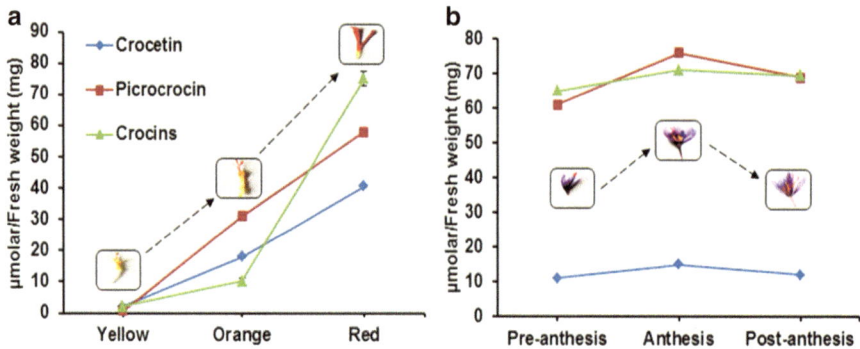

Fig. 1.4 Stage-specific accumulation of apocarotenoids in *C. sativus* (**a**) *yellow*, *orange*, and *red* stages (**b**) pre-anthesis to post-anthesis of stigma development

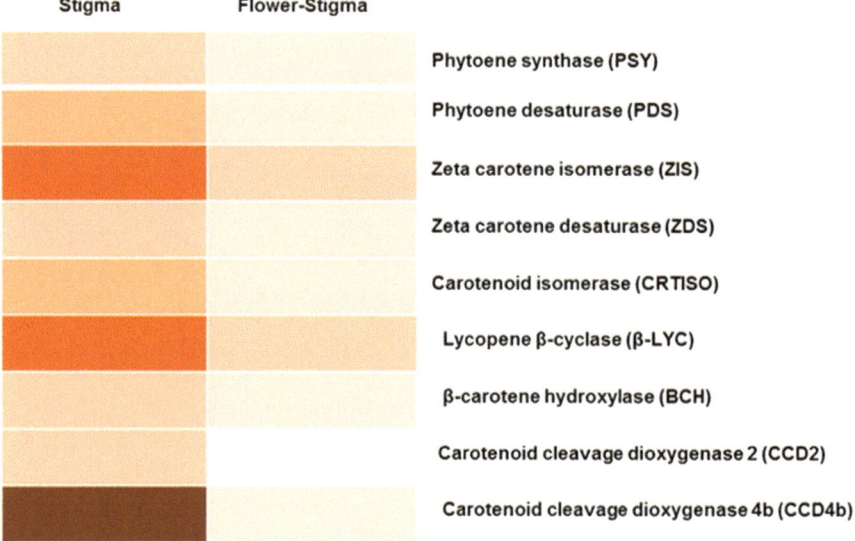

Fig. 1.5 Heat map showing differential expression of apocarotenoid biosynthesis in stigma and flower minus stigma

which are involved in converting phytoene into lycopene are also upregulated in stigma. Further, it is quite interesting that lycopene β-LCY and BCH are highly expressed in stigma, therefore increasing the metabolic flux toward production of zeaxanthin which is then acted upon by CCDs leading to the stigma-specific accumulation of apocarotenoids (Fig. 1.5) (Baba et al. 2015b).

Although there is enough information about accumulation of carotenoids in specialized structures, little is known about the accumulation of their cleavage derivatives (apocarotenoids) (Camara et al. 1995). In *Crocus* style, direct contacts have been observed between chromoplasts and the central vacuole, suggesting that the apocarotenoids might be transported to the vacuole. Such connections between plastids and vacuoles are very uncommon and have been observed in only a few species including leaves of tomato (Crotty and Ledbetter 1973) and *Hypoestes sanguinolenta* (Vaughn and Duke 1981). Electron micrographs of *C. sativus* stigma revealed that carotenoid-derived metabolites got accumulated in vacuoles. This is further supported by the reports that cell suspension cultures of Crocus possess the ability to glycosylate exogenous crocetin and sequester the resulting glycosides in cell vacuoles (Dufresne et al. 1999). Moreover, the accumulation of some apocarotenoids such as mycorradicin is induced by arbuscular mycorrhizal fungi in the vacuole root cells (Klingner et al. 1995; Walter et al. 2000).

Plant secondary metabolites are generally accumulated in specific tissues and developmental stages to carry out different functions. However, these secondary metabolites are often transported from the tissue where they are synthesized to other parts of the plant. Some metabolites are synthesized in the shoot and transported to roots to perform their functions and vice versa. The transport of secondary metabolites has been studied in many plants, and involvement of transporters, mainly ATP-binding cassette (ABC) transporters, has been predicted (Yazaki 2005). In leaves of *Nicotiana plumbaginifolia*, a unique plasma membrane-localized ABC transporter has been reported to be involved in the efflux of endogenous antifungal diterpene, sclareol (Jasinski et al. 2001). The orthologs of this transporter have also been reported from *Arabidopsis* (Campbell et al. 2003) and *Spirodela polyrhiza* (van den Brûle et al. 2002). However, the transport of secondary metabolites may also be due to the process of diffusion as reported in literature, for instance, tomato and some *Arabidopsis* monoterpenes are emitted by leaves by simple diffusion (Aharoni et al. 2003; Lücker et al. 2004; Ohara et al. 2003). The studies directed toward the elucidation of mechanism of transport of carotenoids and their apocarotenoid derivatives are scanty. Among the apocarotenoids, the transport of abscisic acid (ABA) has been studied in detail. It has been reported that ABA transporters including G25 and G40 (AtABCG25 and AtABCG40) (Kuromori et al. 2010; Kang et al. 2010; Verrier et al. 2008) act as ABA transporters in *Arabidopsis*. Kanno et al. (2012) reported that a nitrate transporter (NRT1.2) to be involved in the transport of ABA (Kanno et al. 2012). Another novel apocarotenoid hormone, strigolactone, plays a key role in the initiation of arbuscular mycorrhizal regulation of plant shoot architecture and was recently been reported to be transported by ABC transporter PDR1 in *Petunia hybrida*. It has been observed that *P. hybrida* pdr1 mutants are defective in strigolactone exudation and also exhibit enhanced shoot branching indicating impaired strigolactone allocation (Kretzschmar et al. 2012). The transport of *C. sativus* apocarotenoids from the stigmata to the underground corm during senescence has also been studied (Rubio-Moraga et al. 2010). The decrease in the concentration of Crocins in stigma after anthesis reported earlier (Rubio et al. 2008; Bouvier et al. 2003) was due to their transport from the senescent stigma of the flower to the ovaries and the new developing corms. After being transported to the developing corm, the crocins are deglucosylated resulting in the production of crocetin and are then not detected during the advanced stages of corm development. These studies suggest that the apocarotenoids are absorbed during the early and dynamic phases of corm development, where the glucose molecules rereleased from crocins might help cell initiation and elongation (Rubio-Moraga et al. 2010). Moreover, the remobilizing of these apocarotenoids allows the plant to shed its flowers while evading loss of their nutrients. It should be however noted that these are only the pioneering studies; mechanism of transport of *C. sativus* apocarotenoids is yet to be fully elucidated. The identification of the transporters involved in the transport of these apocarotenoids to the developing corm and their exact roles are expected to open upon new avenues in understanding carotenoid and apocarotenoid metabolism.

1.5   Conclusion

In the last few years, there has been tremendous progress in our knowledge of how carotenoid pathway supplies precursors for the biosynthesis of apocarotenoids in *C. sativus*. A number of studies carried out in the recent past have enormously contributed to decipher the rate-limiting steps, and with recent advancements in research, we are now achieving remarkable progress in the identification of mechanisms that control the flux through these two interlinked pathways. Although apocarotenoid biosynthesis in *C. sativus* has elucidated to a greater extent, there are still some missing links in the pathway. Conversion of picrocrocin to safranal is an enzyme-catalyzed reaction, or a nonenzymatic simple hydrolysis still needs to be verified. The glucosyltransferases involved in the glucosylation of 3-OH-β-cyclocitral are yet to be identified. The knowledge related to regulation and transport of apocarotenoids is still fragmentary. We believed that forthcoming work will be helpful to unveil the different aspects of regulation of apocarotenoid biosynthesis and the specific transporters involved in their transport from stigma to other parts of the plant. In the future, we will be also able to understand how apocarotenoids are synthesized in response to developmental cues and environmental stimuli.

References

Aharoni A, Giri AP, Deuerlein S, Griepink F, de Kogel W-J, Verstappen FWA, Verhoeven HA, Jongsma MA, Schwab W, Bouwmeester HJ (2003) Terpenoid metabolism in wild-type and transgenic Arabidopsis plants. Plant Cell 15:2866–2884

Ahraze O, Rubio-Moraga A, López RC, Gómez-Gómez L (2009) The expression of a chromoplast-specific lycopene beta cyclase gene is involved in the high production of saffron's apocarotenoid precursors. J Exp Bot 61:105–119

Alonso GL, Salinas MR, Garijo J, Sánchez-Fernández MA (2001) Composition of crocins and picrocrocin from Spanish saffron (Crocus sativus L.). J Food Qual 24:219–233

Ashraf N, Jain D, Vishwakarma RA (2015) Identification, cloning and characterization of an ultrapetala transcription factor CsULT1 from Crocus: a novel regulator of apocarotenoid biosynthesis. BMC plant biology 15:25

Auldridge ME, McCarty DR, Klee HJ (2006a) Plant carotenoid cleavage oxygenases and their apocarotenoid products. Curr Opin Plant Biol 9:315–321

Auldridge ME, Block A, Vogel JT, Dabney-Smith C, Mila I, Bouzayen M, Magallanes-Lundback M, DellaPenna D, McCarty DR, Klee HJ (2006b) Characterization of three members of the Arabidopsis carotenoid cleavage dioxygenase family demonstrates the divergent roles of this multifunctional enzyme family. Plant J 45:982–993

Azuma H et al (2002) Floral scent chemistry of mangrove plants. J Plant Res 115:47–53

Baba SA, Malik AH, Wani ZA, Mohiuddin T, Shah Z, Abbas N, Ashraf N (2015a) Phytochemical analysis and antioxidant activity of different tissue types of *Crocus sativus* and oxidative stress alleviating potential of saffron extract in plants, bacteria, and yeast. S Afr J Bot 31(99):80–87

Baba SA, Mohiuddin T, Basu S, Swarnkar MK, Malik AH, Wani ZA, Abbas N, Singh AK, Ashraf N (2015b) Comprehensive transcriptome analysis of *Crocus sativus* for discovery and expression of genes involved in apocarotenoid biosynthesis. BMC Genomics 16(1):1

Baba SA, Jain D, Abbas N, Ashraf N (2015c) Overexpression of Crocus carotenoid cleavage dioxygenase, CsCCD4b, in Arabidopsis imparts tolerance to dehydration, salt and oxidative stresses by modulating ROS machinery. J Plant Physiol 189:114–125

Bouvier F, Suire C, Mutterer J, Camara B (2003) Oxidative remodeling of chromoplast carotenoids: identification of the carotenoid dioxygenase CsCCD and CsZCD genes involved in Crocus secondary metabolite biogenesis. Plant Cell 15:47–62

Bouvier F, Isner JC, Dogbo O, Camara B (2005) Oxidative tailoring of carotenoids: a prospective towards novel functions in plants. Trends Plant Sci 10:187–194

Brandi F, Bar E, Mourgues F, Horváth G, Turcsi E, Giuliano G, Liverani A, Tartarini S, Lewinshon E, Rosati C (2011) Study of 'RedHaven' peach and its white-fleshed mutant suggests a key role of CCD4 carotenoid dioxygenase in carotenoid and norisoprenoid volatile metabolism. BMC Plant Biol 11:1–24

Britton G, Liaaen-Jensen S, Pfander H (eds) (2009) Carotenoids volume 5: nutrition and health, vol 5. Springer Science & Business Media

Bugg TDH, Ahmad M, Hardiman EM, Rahmanpour R (2011) Nat Prod Rep 28:1883–1896

Camara B, Hugueney P, Bouvier F, Kuntz M, Monéger R (1995) Biochemistry and molecular biology of chromoplasts development. Int Rev Cytol 163:175–247

Campbell EJ, Schenk PM, Kazan K, Penninckx IA, Anderson JP, Maclean DJ, Cammue BP, Ebert PR, Manners JM (2003) Pathogen-responsive expression of a putative ATP-binding cassette transporter gene conferring resistance to the diterpenoid sclareol is regulated by multiple defense signaling pathways in Arabidopsis. Plant Physiol 133:1272–1284

Carmona M, Zalacain A, Sánchez A, Novella J, Alonso G (2006) Crocetin esters, picrocrocin and its related compounds present in Crocus sativus stigmas and Gardenia jasminoides fruits. Tentative identification of seven new compounds by LC–ESI-MS. J Agric Food Chem 54:973

Castillo R, Fernández JA, Gómez-Gómez L (2005) Implications of carotenoid biosynthetic genes in apocarotenoid formation during the stigma development of Crocus sativus and its closer relatives. Plant Physiol 139(2):674–689

Cazzonelli CI et al (2009) Regulation of carotenoid composition and shoot branching in Arabidopsis by a chromatin modifying histone methyltransferase, SDG8. Plant Cell 21(1):39–53

Cazzonelli CI, Pogson BJ (2010) Source to sink: regulation of carotenoid biosynthesis in plants. Trends Plant Sci 15(5):266–274

Chaudhary N et al (2010) Carotenoid biosynthesis genes in rice: structural analysis, genome-wide expression profiling and phylogenetic analysis. Mol Genet Genomics 283:13–33

Corona V et al (1996) Regulation of a carotenoid biosynthesis gene promoter during plant development. Plant J 9:505–512

Cote F, Cormier F, Dufresne C et al (2001) A highly specific glucosyltransferase is involved in the synthesis of crocetin glucosylesters in Crocus sativus cultured cells. J Plant Physiol 158:553–560

Crotty WJ, Ledbetter MC (1973) Membrane continuities involving chloroplasts and other organelles in plant cells. Science 182(4114):839–841

Curro P, Lanuza F, Micali G (1986) Valutazione Della frazione volatile dello zafferano mediante gascromatografia dello spazio di testa. Rass Chimica 6:331–334

Davies KM (2007) Genetic modification of plant metabolism for human health benefits. Mutat Res 622:122–137

DellaPenna D, Pogson BJ (2006) Vitamin synthesis in plants: tocopherols and carotenoids. Annu Rev Plant Biol 57:711–738

Dhingra V, Seshadri T, Mukerjee S (1975) Minor carotenoid glycosides from saffron (Crocus sativus). Ind J Chem 13:339–341

Donaldson JMI et al (1990) Floral attractants for the Cetoniinae and Rutelinae (Coleoptera: Scarabaeidae). J Econ Entomol 83:1298–1305

Dong H et al (2007) The Arabidopsis spontaneous cell death1 gene, encoding a zeta-carotene desaturase essential for carotenoid biosynthesis, is involved in chloroplast development, photo-protection and retrograde signalling. Cell Res 17:458–470

Dufresne C, Cormier F, Dorion S, Niggli UA, Pfister S, Pfander H (1999) Glycosylation of encapsulated crocetin by a Crocus sativus L. cell culture. Enzym Microb Technol 24(8):453–462

Eroglu A, Harrison EH (2013) Carotenoid metabolism in mammals, including man: formation, occurrence, and function of apocarotenoids. J Lipid Res 54:1719–1730

Fang J et al (2008) Mutations of genes in synthesis of the carotenoid precursors of ABA lead to pre-harvest sprouting and photo-oxidation in rice. Plant J 54:177–189

Farre G et al. (2010) Travel advice on the road to carotenoids in plants. Plant Sci 179:28–48

Farré-Armengol G, Filella I, Llusia J, Peñuelas J (2013) Floral volatile organic compounds: between attraction and deterrence of visitors under global change. Perspect Plant Ecol Evol Syst 15(1):56–67

Flath RA et al (1994) Alpha-ionol as attractant for trapping Batrocera latifrons (Diptera: Tephritidae). J Econ Entomol 87:1470–1476

Fraser PD, Bramley PM (2004) The biosynthesis and nutritional uses of carotenoids. Prog Lipid Res 43:228–265

Frusciante S, Diretto G, Bruno M, Ferrante P, Pietrella M, Prado-Cabrero A et al (2014) Novel carotenoid cleavage dioxygenase catalyzes the first dedicated step in saffron crocin biosynthesis. Proc Natl Acad Sci U S A 111:12246–12251

Gomez-Roldan V, Fermas S, Brewer PB, Puech-Pages V, Dun EA, Pillot JP, Letisse F, Matusova R, Danoun S, Portais JC, Nouwmeester H, Becard G, Beveridge CA, Huang FC, Molnár P, Schwab W (2009) Cloning and functional characterization of carotenoid cleavage dioxygenase 4 genes. J Exp Bot 60:3011–3022

Gonzalez-Jorge S, Ha S, Magallanes-Lundback M, Gilliland LU, Zhou A, Lipka AE, Nguyen YN, Angelovici R, Lin H, Cepela J, Little H, hBuell CR, Gore MA, Della-Penna D (2013) Carotenoid cleavage dioxygenase 4 is a negative regulator of b-carotene content in Arabidopsis seeds. Plant Cell 25:4812–4826

Gregory MJ, Menary RC, Davies NW (2005) Effect of drying temperature and air flow on the production and retention of secondary metabolites in saffron. J Agric Food Chem 53:5969–5975

Halliwell B (2006) Oxidative stress and neurodegeneration; where are we now? J Neurochem 97:1634–1658

Halliwell B (2007) Oxidative stress and cancer: have we moved forward? Biochem J 401:1–11

Himeno H, Sano K (1987) Synthesis of crocin, picrocrocin and safranal by saffron stigma-like structures proliferated in vitro. Agric Biol Chem 9(51):2395–2400

Howitt CA et al (2009) Alternative splicing, activation of cryptic exons and amino acid substitutions in carotenoid biosynthetic genes are associated with lutein accumulation in wheat endosperm. Funct Integr Genomics 9:363–376

Isaacson T et al (2002) Cloning of tangerine from tomato reveals a carotenoid isomerase essential for the production of beta-carotene and xanthophylls in plants. Plant Cell 14:333–342

Isaacson T et al (2004) Analysis in vitro of the enzyme CRTISO establishes a poly-cis-carotenoid biosynthesis pathway in plants. Plant Physiol 136:4246–4255

Jasinski M, Stukkens Y, Degand H, Purnelle B, Marchand-Brynaert J, Boutry M (2001) A plant plasma membrane ATP binding cassette-type transporter is involved in antifungal terpenoid secretion. Plant Cell 13:1095–1107

Kamoda S, Saburi Y (1993) Biotechnol Biochem 57:926–930

Kang J et al (2010) PDR-type ABC transporter mediates cellular uptake of the phytohormone abscisic acid. Proc Natl Acad Sci U S A 107:2355–2360

Kanno Y et al (2012) Identification of an abscisic acid transporter by functional screening using the receptor complex as a sensor. Proc Natl Acad Sci U S A 109:9653–9658

Khun R, Winterstein A (1934) Die Dihydroverbindung der isomeren Bixine und die Elektronen-Konfiguration der Polyene. Ber Dtsch Chem Ges 67:344–347

Klingner A, Bothe H, Wray V, Marner FJ (1995) Identification of a yellow pigment formed in maize roots upon mycorrhizal colonization. Phytochemistry 38:53–55

Kloer DP, Schulz GE (2006) Structural and biological aspects of carotenoid cleavage. Cell Mol Life Sci CMLS 63(19–20):2291–2303

Kretzschmar T, Kohlen W, Sasse J, Borghi L, Schlegel M, Bachelier JB, Reinhardt D, Bours R, Bouwmeester HJ, Martinoia E (2012) A petunia ABC protein controls strigolactone-dependent symbiotic signalling and branching. Nature 483(7389):341–344

Krinsky NI, Johnson EJ (2005) Carotenoid actions and their relation to health and disease. Mol Asp Med 26:459–516

Kuromori T et al (2010) ABC transporter At ABCG25 is involved in abscisic acid transport and responses. Proc Natl Acad Sci U S A 107:2361–2366

Li N, Lin G, Kwan YW, Min ZD (1999) Simultaneous quantification of five major biologically active ingredients of saffron by high-performance liquid chromatography. J Chromatogr A 849:349–355

Li F et al (2007) Maize Y9 encodes a product essential for 15-cis-zeta- carotene isomerization. Plant Physiol 144:1181–1189

Li F et al (2008) PSY3, a new member of the phytoene synthase gene family conserved in the Poaceae and regulator of abiotic stress induced root carotenogenesis. Plant Physiol 146:1333–1345

Liao YH, Houghton PJ, Hoult JRS (1999) Novel and known constituents from Buddleja species and their activity against leukocyte eicosanoid generation. J Nat Prod 62(9):1241–1245

Lücker J, Schwab W, Franssen MC, Van Der Plas LH, Bouwmeester HJ, Verhoeven HA (2004) Metabolic engineering of monoterpene biosynthesis: two-step production of (+)-transisopiperitenol by tobacco. Plant J 39:135–145

Ma J, Li J, Zhao J, Zhou H, Ren F, Wang L, Gu C, Liao L, Han Y (2014) Inactivation of a gene encoding carotenoid cleavage dioxygenase (CCD4) leads to carotenoid-based yellow coloration of fruit flesh and leaf midvein in peach. Plant Mol Biol Rep 32:246–257

Maggi L, Carmona M, del Campo CP, Kanakis CD, Anastasaki E, Tarantilis PA (2009) Worldwide market screening of saffron volatile composition. J Sci Food Agric 89:1950–1954

McQuate GT, Peck SL (2001) Enhancement of attraction of alpha-ionol to male Bactrocera latifrons (Diptera: Tephritidae) by addition of a synergist, cade oil. J Econ Entomol 94:39–46

Messing SAJ, Gabelli SB, Echeverria I, Vogel JT, Guan JC, Tan BC, Klee HJ, McCarty DL, Amzel LM (2010) Structural insights into maize viviparous14, a key enzyme in the biosynthesis of the phytohormone abscisic acid. Plant Cell 22(9):2970–2980

Moiseyev G, TY, Chen Y, Gentleman S, Redmond TM, Crouch RK, Ma J (2006) J Biol Chem 281:2835–2840

Moraga AR, Nohales PF, Pérez JAF, Gómez-Gómez L (2004) Glucosylation of the saffron apocarotenoid crocetin by a glucosyltransferase isolated from Crocus sativus stigmas. Planta 219(6):955–966

Namin MH, Ebrahimzadeh H, Ghareyazie B, Radjabian T, Gharavi S, Tafreshi N (2009) In vitro expression of apocarotenoid genes in Crocus sativus L. Afr J Biotechnol 8(20):5378–5382

Oberhauser V, Voolstra O, Bangert A, Von Lintig J, Vogt JK (2008) NinaB combines carotenoid oxygenase and retinoid isomerase activity in a single polypeptide. Proc Natl Acad Sci 105(48):19000–19005

Ohara K, Ujihara T, Endo T, Sato F, Yazaki K (2003) Limonene production in tobacco with Perilla limonene synthase cDNA. J Exp Bot 54:2635–2642

Park H et al (2002) Identification of the carotenoid isomerase provides insight into carotenoid biosynthesis, prolamellar body formation, and photomorphogenesis. Plant Cell 14:321–332

Pasare SA, Ducreux LJM, Morris WL, Campbell R, Sharma SK, Roumeliotis E, Kohlen W, van der Krol S, Bramley PM, Roberts AG, Fraser PD, Taylor MA (2013) The role of the potato (Solanum tuberosum) CCD8 gene in stolon and tuber development. New Phytol 198:1108–1120

Pfander H, Wittwer F (1975) Carotenoid composition in safran. Helv Chim Acta 58:2233–2236

Pfister S, Meyer P, Steck A, Pfander H (1996) Isolation and structure elucidation of carotenoid-glycosyl esters in Gardenia fruits (Gardenia jasminoides Ellis) and saffron (Crocus sativus Linne). J Agric Food Chem 44:2612–2615

Poliakov E, gentleman S, Cunningham FX jr, Miller-lhli Nj, Remond TM (2005) Key role of conserved histidine in recombinant mouse beta-carotene 15 15 monooxygenase-1. JBC 280:29217–29223

Raina BL, Agarwal SG, Bhatia AK, Gaur GS (1996) Changes in pigments and volatiles of saffron (Crocus sativus L.) during processing and storage. J Sci Food Agric 71:27–32

Rao AV, Rao LG (2007) Carotenoids and human health. Pharmacol Res 55(3):207–216

Rubio A, Rambla JL, Santaella M, Gomez MD, Orzaez D, Granell A, Gómez-Gómez L (2008) Cytosolic and plastoglobule-targeted carotenoid dioxygenases from Crocus sativus are both involved in beta-ionone release. J Biol Chem 283:24816–24825

Rubio-Moraga A, Trapero A, Ahrazem O, Gómez-Gómez L (2010) Crocins transport in Crocus sativus: the long road from a senescent stigma to a newborn corm. Phytochemistry 71(13):1506–1513

Rubio-Moraga A, Ahrazem O, Pérez-Clemente RM, Gómez-Cadenas A, Yoneyama K, López-Ráez JA, Molina RV, Gómez-Gómez L (2014) Apical dominance in saffron and the involvement of the branching enzymes CCD7 and CCD8 in the control of bud sprouting. BMC Plant Biol 14(1):171

Sauter A, Davies WJ, Hartung W (2001) The long-distance abscisic acid signal in the droughted plant: the fate of the hormone on its way from root to shoot. J Exp Bot 52:1991–1997

Schwartz SH, Qin XQ, Loewen MC (2004) The biochemical characterization of two carotenoid cleavage enzymes form Arabidopsis indicates that a carotenoid-derived compound inhibits lateral branching. J Biol Chem 279:46940–46945

Simkin AJ, Schwartz SH, Auldridge M, Taylor MG, Klee HJ (2004a) The tomato carotenoid cleavage dioxygenase 1 genes contribute to the formation of the flavour volatiles β-ionone, pseudoionone and geranylacetone. Plant J 40:882–892

Simkin AJ, Underwood BA, Auldridge M, Loucas HM, Shibuya K, Schmelz E, Clark DG, Klee HJ (2004b) Circadian regulation of the PhCCD1 carotenoid cleavage dioxygenase controls emission of b-ionone, a fragrance volatile of Petunia flowers. Plant Physiol 136:3504–3514

Speranza G, Dada G, Manitto P, Monti D, Gramatica P (1984) 13-cis-crocin: a new crocinoid of saffron. Gazz Chim Ital 114:189–192

Sui X, Kiser PD, Lintig JV, Palczewski K (2013) Structural basis of carotenoid cleavage: from bacteria to mammals. Arch Biochem Biophys 539(2):203–213

Tan BC, Schwartz SH, Zeevaart JAD, McCarty DR (1997) Proc Natl Acad Sci 94:12235–12240

Tarantilis PA, Tsoupras G, Polissiou M (1995) Determination of saffron (Crocus sativus L.) components in crude plant extract using high-performance liquid chromatography–UV–visible photodiode-array detection-mass spectrometry. J Chromatogr A 699:107–118

Toledo-Ortiz G, Huq E, Rodríguez-Concepción M (2010) Direct regulation of phytoene synthase gene expression and carotenoid biosynthesis by phytochrome-interacting factors. PNAS 107:11626–11631

Umehara M, Hanada A, Yoshida S, Akiyama K, Arite T, Takeda-Kamiya N, Magome H, Kamiya Y, Shirasu K, Yoneyama K, Kyozuka J, Yamaguchi S (2008) Inhibition of shoot branching by new terpenoid plant hormones. Nature 455:195–200

van den Brûle S, Müller A, Fleming AJ, Smart CC (2002) The ABC transporter SpTUR2 confers resistance to the antifungal diterpene sclareol. Plant J 30:649–662

Vaughn KC, Duke SO (1981) Evaginations from the plastid envelope: a method for transfer of substances from plastid ton vacuole. Cytobios 32:89–95

Verma RS, Middha D (2010) Analysis of saffron (Crocus sativus L. stigma) components by LC–MS–MS. Chromatographia 71(1–2):117–123

Verrier PJ et al (2008) Plant ABC proteins-a unified nomenclature and updated inventory. Trends Plant Sci 13:151–159

Vom Endt D, Kijne JW, Memelink J (2002) Transcription factors controlling plant secondary metabolism: what regulates the regulators? Phytochemistry 61(2):107–114

Walter MH, Strack D (2011) Carotenoids and their cleavage products: biosynthesis and functions. Nat Prod Rep 28:663–692

Walter MH, Fester T, Strack D (2000) Arbuscular mycorrhizal fungi induce the non-mevalonate methylerythritol phosphate pathway of isoprenoid biosynthesis correlated with accumulation of the "yellow pigment" and other apocarotenoids. Plant J 21:571–578

Walter MH, Floss DS, Strack D (2010) Apocarotenoids: hormones, mycorrhizal metabolites and aroma volatiles. Planta 232:1–17

Wang SC, Tseng TY, Huang CM, Tsai TH (2004) Gardenia herbal active constituents: applicable separation procedures. J Chromatogr B 812:193–202

Wei S, Hannoufa A, Soroka J, Xu N, Li X, Zebarjadi A, Gruber M (2011) Enhanced β-ionone emission in Arabidopsis over-expressing AtCCD1 reduces feeding damage in vivo by the crucifer flea beetle. Environ Entomol 40(6):1622–1630

Welsch R, Maass D, Voegel T, DellaPenna D, Beyer P (2007) Transcription factor RAP2.2 and its interacting partner SINAT2: stable elements in the carotenogenesis of Arabidopsis leaves. Plant Physiol 145:1073–1085

Welsch R et al (2008) A third phytoene synthase is devoted to abiotic stress-induced abscisic acid formation in rice and defines functional diversification of phytoene synthase genes. Plant Physiol 147:367–380

Winterstein E, Teleczky J (1922) Constituents of the saffron. I. Picrocrocin. Helv Chimica Acta 5:376–400

Yang B, Guo Z, Liu R (2005) Crocin synthesis mechanism in Crocus sativus. Tsinghua Sci Technol 10(5):567–572

Yazaki K (2005) Transporters of secondary metabolites. Curr Opin Plant Biol 8(3):301–307

Yu F et al (2007) Variegation mutants and mechanisms of chloroplast biogenesis. Plant Cell Environ 30:350–365

Chapter 2
Carotenoid Cleavage Dioxygenases
of *Crocus sativus* L.

Abstract Carotenoid cleavage dioxygenases (CCDs) form a multienzyme family, the members of which are involved in the production of a diversity of apocarotenoids. The apocarotenoid module vital physiological and developmental processes in plants. This chapter deals with the different aspects of plant CCDs in general and *C. sativus* in particular such as structure and reaction mechanisms. Further, this chapter also discusses the role of CCDs in plants and their application in plant biotechnology.

Keywords Carotenoid cleavage dioxygenase • Apocarotenoid • Saffron

2.1 Introduction

Carotenoids are C_{40} isoprenoids and form one of the most diverse classes of natural compounds. They serve multitude of functions ranging from photosynthetic pigments to nutritional compounds and antioxidants. Carotenoids undergo a reaction involving incorporation of both oxygen atoms from molecular oxygen into their substrates across a double bond resulting in the production of two ketone- or aldehyde-containing cleavage products (Auldridge et al. 2006a, b). These cleavage products are known as apocarotenoids (Table 2.1). They are widespread in plant kingdom and carry out vital physiological and developmental processes in plants. They act as signaling molecules, growth regulators, pollinator attractants, and predator repellents and also provide some plants with competitive advantage over others (Bouvier et al. 2005a, b). In recent years, a number of enzymes involved in the production of apocarotenoids have been identified and characterized across plant kingdom. All of these belong to a family of enzymes called carotenoid cleavage dioxygenases (CCDs). These enzymes are promiscuous as far as their substrate specificity is concerned but show high regio- and stereospecificity for the double bonds they cleave (Huang et al. 2009).

The first enzyme found to precisely cleave carotenoids, viviparous14 (VP14), was recognized by the investigation of viviparous abscisic acid-deficient mutant of maize

© Springer Science+Business Media Singapore 2016   23
S.A. Baba, N. Ashraf, *Apocarotenoids of Crocus sativus L: From biosynthesis to pharmacology*, SpringerBriefs in Plant Science, DOI 10.1007/978-981-10-1899-2_2

Table 2.1 Some important plant CCDs and their apocarotenoids

Name of the plant	CCD	Cleavage products
Bixa	*Bo*CCD4, *Bo*CCD1	Bixin, (9′Z)-apo-6′-lycopenoate, methyl (9′Z)-6′-oxo 6,6′-diapocarotenoate, and methyl (all-E)-8′-apo-b-caroten-8′-oate
Crocus sativus	*Cs*CCd4a, *Cs*CCD4b	β-Ionone
	*Cs*CCD1	Crocin
	*Cs*CCD2	
Chrysanthemum morifolium	*Cm*CCD4a, *Cm*CCD4b	β-Ionone
Petunia hybrida	*Ph*CCD1	β-Ionone
Lycopersicon esculentum	*Le*CCD1a, *Le*CCD1b	Geranylacetone
		Pseudoionone
		α- and β-ionone
Vitis vinifera	*Vv*CCD1, VvCCD4,	Pseudoionone, β-ionone, 3-hydroxy-β-ionone, geranylacetone
Rosa damascena	*Rd*CCD4	β-ionone, grasshopper ketone
Cucumis melo	*Cm*CCD1	Grasshopper ketone
		Geranylacetone
		Pseudoionone
		α- and β-ionone
Osmanthus fragrans	*Of*CCD1	α-Ionone, β-ionone
Fragaria ananassa	*Fa*CCD1	3-hydroxy-β-ionone, 3-hydroxy-α-ionone
Malus domestica	*Md*CCD4	β-Ionone
Arabidopsis thaliana	*At*CCD4, *At*CCD7, *At*CCD8	β-ionone, strigolactones

(Schwartz et al. 1997; Tan et al. 2003). This pioneering work facilitated the discovery of related enzymes in different plant species. The CCD multienzyme family is generally characterized by an amphipathic α-helix domain, a catalytic domain which contains four conserved histidine residues responsible for nonheme iron coordination, Fe^{2+} as a cofactor, and a plastid-targeting transit peptide except for some CCDs that are localized in the cytoplasm. Based on the sequence homology, two types of carotenoid dioxygenases have been reported in higher plants which include 9-*cis*-carotenoid cleavage dioxygenases (NCEDs) and carotenoid cleavage dioxygenases (CCDs). The CCD family of *Arabidopsis thaliana* consists of nine members which include five NCEDs that are involved in the biosynthesis of the plant hormone ABA and four CCDs that are involved in various carotenoid cleavage reactions.

Recently, the CCD subfamily has been further expanded to CCD1, CCD4, CCD7, and CCD8 groups (Fig. 2.1) based on substrate specificities and cleavage site (Auldridge et al. 2006a, b). CCD1 enzymes have wide substrate and cleavage site specificity (Bouvier et al. 2005a, b; Auldridge et al. 2006a, b). They cleave cyclic and acyclic all-*trans* carotenoids, as well as apocarotenoids, like β-apo-8′-carotenal, β-apo-10′-carotenal, and apolycopenals leading to the production of

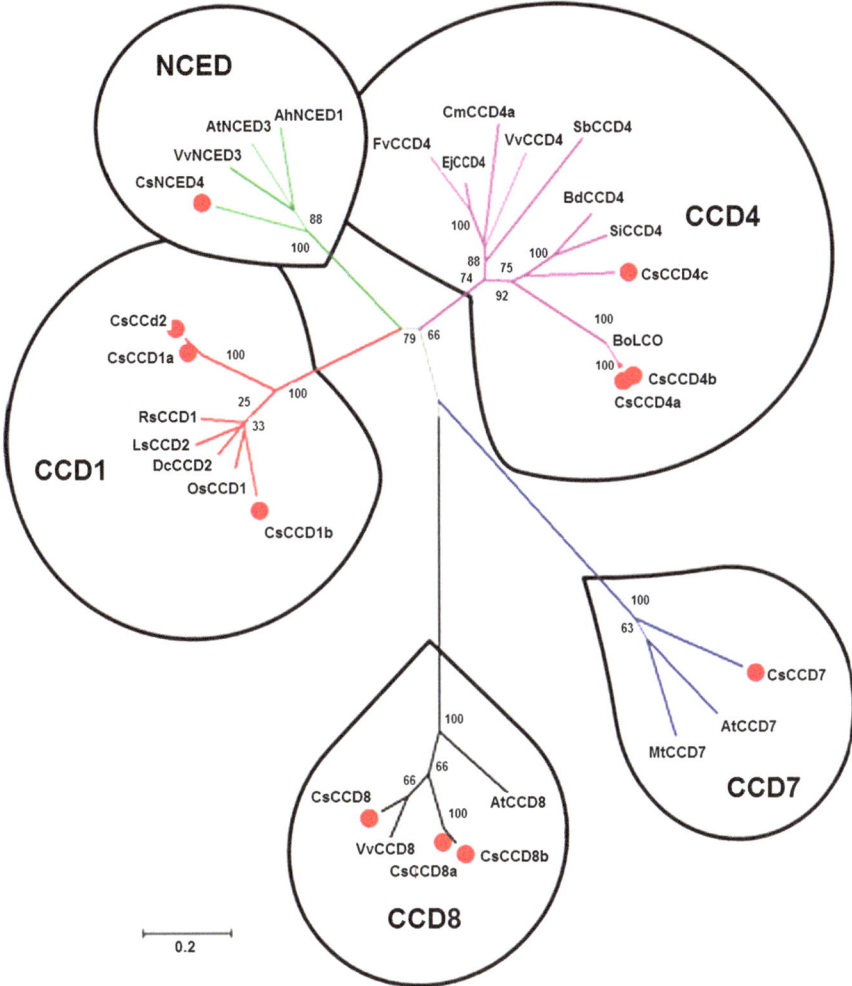

Fig. 2.1 Phylogenetic analysis of *Crocus sativus* and other plant CCDs. The sequences in this study include the following: *C. sativus Cs*CCD1a (CAC95131.1), *Cs*CCD1b (CAC95132.1), *Cs*CCD4a (ACD62476.1), *Cs*CCD4b (EU523663.1), *Cs*CCD4c (JN131499.1), *CsZ*CD (AJ489276.1), *Cs*CCD2 (EU523661.1), *Cs*CCD8a (A0A075IBX5), *Cs*CCD8b (A0A075IGQ2), *Cs*CCD7 (A0A075IEG5), and *Cs*NCED (B7SNW4); *Brachypodium distachyon Bd*CCD4 (XM 003570134.1); *Setaria italica* SiCCD4 (XM 004953541.1); *Bixa orellana Bo*LCO (AJ489277.1); *Fragaria vesca Fv*CCD4 (XM 004297596.1); *Vitis vinifera Vv*CCD4 (JQ712828.1), VvCCD8 (002281203.1), and *Vv*NCED1 (AY337613.1); *Scutellaria baicalensis Sb*CCD4 (KC760148.1); *Chrysanthemum x morifolium Sm*CCD4a (EU334432.1); *Eriobotrya japonica Ej*CCD4 (JQ317782.1); *Arabidopsis thaliana At*CCD7 (NM 130064.4), *At*CD8 (NM 119434.3), and *At*NCED3 (NM 112304.2); *Citrus sinensis Sci*CCD8 (006476067.1); *Malus toringoides Mt*CCD7 (KF887976.1); *Arachis hypogaea Ah*NCED1 (AJ574819.2); *Osmanthus fragrans Of*CCD1 (AB526197.1); *Rosa damascena Rd*CCD1 (EU327776.1); *Daucus carota Dc*CCD1 (DQ192203.1)

various volatiles like β-ionone, 6-methyl-5-hepten-2-one (MHO; C8), and gera-
nial (Schmidt et al. 2006; Vogel et al. 2010; Ilg et al. 2010). CCD4 enzymes are
involved in the production of β-ionone by cleaving β-carotene or β-apo-8'-
carotenal at the C9–C10 double bond (Huang et al. 2009; Rubio et al. 2008).
Similarly, CCD7 and CCD8 are involved in strigolactone biosynthesis. These two
enzymes carry out the sequential cleavage of β-carotene leading to the production
of carlactone, a strigolactone-like compound that has been very recently identified
in planta.

2.2 Carotenoid Cleavage Dioxygenase Family of *Crocus sativus* L.

In *C. sativus* stigma, three apocarotenoids crocin, picrocrocin, and safranal accumulate in significant amounts (Baba et al. 2015b). These apocarotenoids are responsible for the color, flavor, and aroma of saffron, making it one of the world's costliest
spice (Rubio et al. 2008; Baba et al. 2015a). Crocins present in *C. sativus* are power-
ful free radical quenchers and have been used in traditional medicine since thou-
sands of years. They display a variety of health benefits and have gained attention
due to their sedative, analgesic, and anticancer properties (Zhang et al. 2009).
However, limited availability of crocins forms a bottleneck in commercial utiliza-
tion of crocins from saffron and results in the in prohibitive market prices.
Principally, *Crocus* apocarotenoids including crocin are synthesized by the oxidative cleavage of either zeaxanthin or β-carotene by CCDs. From the last one decade,
a number of CCDs have been isolated from *C. sativus* and have been found to be
involved in a range of apocarotenoids. Three isoforms of CCD4 (CsCCD4a,
CsCCD4b, and CCD4c) and two isoforms of CsCCD1 (CsCCD1a and CsCCD1b)
have been identified and have been found to be involved in the 9,10 (9',10') oxidative cleavage of β-carotene leading to the production of β-ionone which is considered as an important signaling molecule during abiotic stress (Rubio et al. 2008;
Baba et al. 2015c). Another enzyme known as zeaxanthin cleavage dioxygenase
(ZCD) has also been isolated from *C. sativus* and was reported to be involved in the
production of crocetin dialdehyde by the oxidative cleavage of zeaxanthin at 7,8
(7',8') double bond. However, it was not until recently that CsZCD was found to be
a truncated form of CsCCD4b and a new carotenoid cleavage dioxygenase CsCCD2
was found to be involved in the production of crocin from zeaxanthin. Most of
CsCCD enzymes are localized in chloroplasts where they get maximum access to
the carotenoid substrates. However, CsCCD1 and CsCCD2 are localized in the
cytoplasm (Rubio et al. 2008; Bouvier et al. 2003). Besides these CCDs, two other
CCDs known as CsCCD7 and CsCCD8 have also been isolated from *C. sativus*.
These enzymes are believed to be involved in the biosynthesis of a novel strigolactone hormone through sequential cleavage of β-carotene. Strigolactones that have
recently been identified from plants are considered important in development of
roots. These molecules are also known to trigger the germination of parasitic plant

seeds and stimulate symbiotic fungi. Phylogenetic analysis of *C. sativus* CCDs (Fig. 2.1) has shown that in CCD subfamily, two major duplications have taken place. These duplications have ultimately led to the appearance of two lineages that evolved into CCD7 and CCD8. Moreover, CCD7 and CCD8 subfamilies had similar evolutionary history but different from t CCD1 and CCD4 subfamilies. Notwithstanding, the five subfamilies of CCDs that form separate clusters in phylogenetic tree identify different cleavage sites of carotenoids (Baba et al. 2015c). However, CsCCD2, a newly identified CCD from *C. sativus* involved in crocin biosynthesis, is an exception to this trend. CsCCD2 clusters in CCD1 subfamily, but identifies a different cleavage site than CsCCD1 (Frusciante et al. 2014; Baba et al. 2015c; Ashraf et al. 2015).

2.3 Structure of Carotenoid Cleavage Dioxygenases

The CCD multienzyme family is generally characterized by an amphipathic a-helix domain, a catalytic domain containing four evolutionary conserved histidine residues required for nonheme iron coordination, Fe^{2+} as a cofactor, and a plastid-targeting transit peptide except for some CCDs that are localized in the cytoplasm like CCD1 (Walter et al. 2000). CCDs cleave linear carotenoids either at the C5–C6, the C7–C8, or the C9–C10 positions and cyclic carotenoids at the C9–C10 position (Vogel et al. 2010; Ilg et al. 2010). All CCDs of plant and animal origin are presumed to act similarly in that they incorporate both oxygen atoms from molecular oxygen into their substrates across a double bond resulting in the production of two aldehyde-containing cleavage products (Auldridge et al. 2006a, b). Recently, the crystal structure of apocarotenoid-15,15′-oxygenase (ACO) from cyanobacterium *Synechocystis* was successfully determined, and this enzyme was found to have seven-bladed beta-propeller tertiary structure. The ACO protein structure showed that the enzyme contains a Fe^{2+} ion as cofactor. The Fe^{2+} ion is coordinated to four conserved histidine residues in the active site of the enzyme. The four-histidine–iron arrangement is conserved CCD multienzyme family. Apart from these features, their amino acid sequences are quite unrelated. The plant CCDs show high degree of similarity with bacterial lignostilbene dioxygenases (LSD) in overall sequence, conservation of histidine residues, iron as a cofactor, and cleavage of double bond suggesting that the enzymes might have evolved from a common ancestor (Kamoda and Saburi 1993; Bugg et al. 2011; tan et al. 1997). Recent studies on some CCDs have shown the presence of intrinsic isomerase activity including RPE65 and NinaB and mammalian CCDs in general (Moiseyev et al. 2006; Oberhauser et al. 2008). However, no intrinsic isomerase activity has been seen in plant CCDs which could be due to the independent presence of iron- dependent carotenoid isomerases. As is the case with other plant CCDs, *C. sativus* also show conservation in the overall structure. *C. sativus* CCDs, such as CsCCD4 isoforms, show the conservation of four His and Glu catalytic residues (Fig. 2.2). Further, the residues which are required for interaction with the substrate are also conserved. These *C. sativus* CCD4 isoforms have a hydrophobic patch for membrane binding that helps them

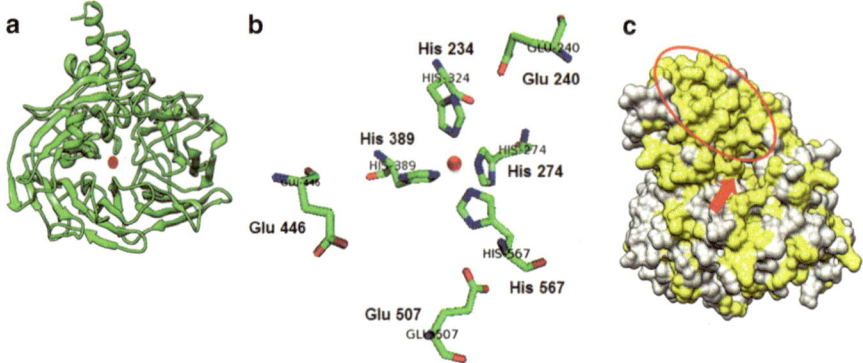

Fig. 2.2 Structure of *C. sativus* CCD4 isoform. (**a**) Three-dimensional model showing seven β-propeller structures. (**b**) Conserved His and Glu residues. (**c**) Surface of CCD4 showing presence of hydrophobic patch

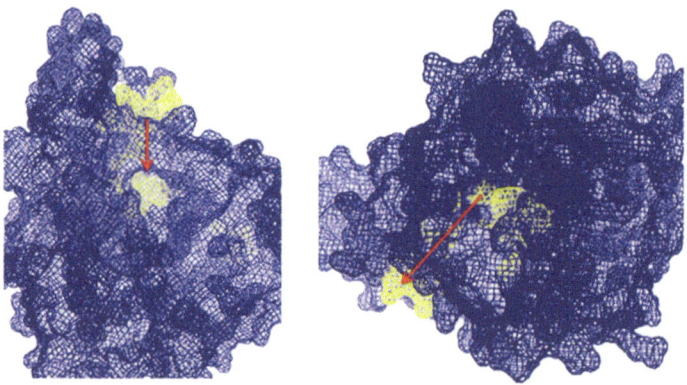

Fig. 2.3 Different views of *C. sativus* CCD4 showing the presence of tunnels (*yellow*) for substrate channeling

extract substrates from plastoglobules present in the stigma of *C. sativus* flowers (Baba et al. 2015c). Another prominent feature of *C. sativus* and other plant CCDs is a tunnel extending from outside of the protein and entering its active center relatively perpendicular to the propeller axis which helps the enzyme to channel the substrates to its active site (Fig. 2.3).

2.4   Reaction Mechanism of Carotenoid Cleavage Dioxygenases: Mono- or Dioxygenase

The reaction mechanism has been studied for a limited number of CCDs which include NCED, mammalian β-carotene dioxygenase, and *A. thaliana* CCD1. All CCD enzymes are iron-dependent oxygenases. Although the requirement of Fe^{2+} and O_2 for the reaction is clear, the nature of the involved oxygen species is yet to be clarified. Since, carotenoids are highly reactive due to the presence of conjugated double bonds. These conjugated doubles make them amenable to unspecific oxygenation and therefore cleavage. This type of cleavage can occur through a number of chemical, thermal, or enzymatic routes. However, the CCD-catalyzed cleavage reactions of carotenoids are well defined. Although controversies still persist regarding the mechanism of CCD-catalyzed reactions, it is clear that the C–C double bond is cleaved to the resultant aldehydes or ketones using dioxygen as reactant. Early recommendations for a dioxygenase mechanism for NCED were built upon $^{18}O_2$ labeling studies. These studies revealed nearly 100% labeling of abscisic acid in the carboxylic acid position (Zeevaart et al. 1989). However, $^{17}O_2$ labeling studies carried out on the mammalian β-carotenoid 15,15′-oxygenase gave only 50% incorporation of ^{17}O into each of the two aldehyde products which was assumed as a monooxygenase mechanism (Leuenberger et al. 2001), involving an epoxide intermediate, shown in Fig. 2.4. This made the mechanism of CCDs somewhat controversial. In the case of the monooxygenase reaction mechanism, an epoxide intermediate is formed at the double bond site with ROS, and only one of the two oxygen molecules participates in the reaction. This mechanism was further supported by an in vitro study carried on chicken BCO1 purified from intestinal mucosa. The cleavage reaction of this purified enzyme was

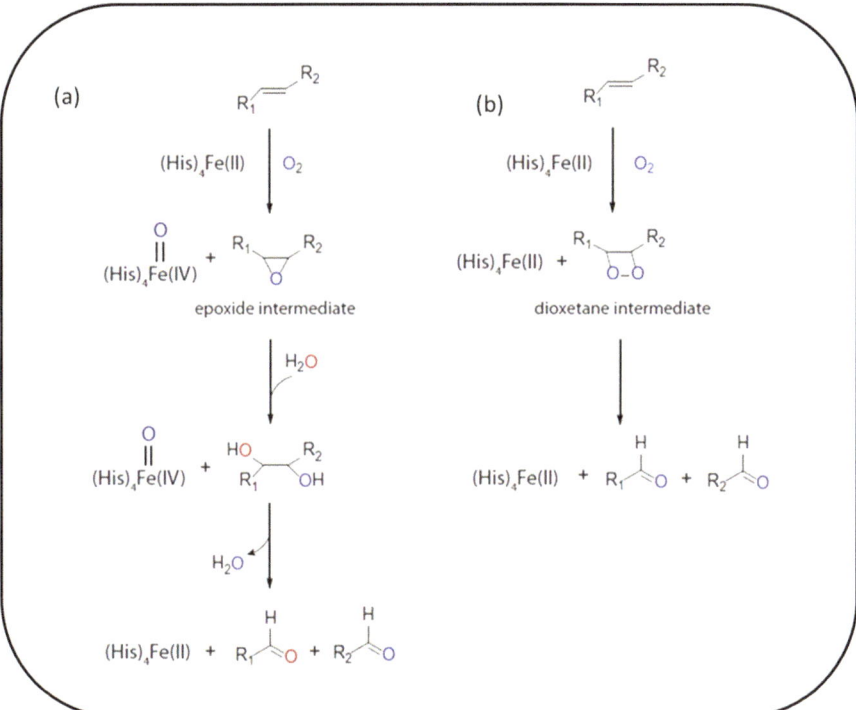

Fig. 2.4 Proposed reaction mechanism for CCDs: (**a**) monooxygenase mechanism, (**b**) dioxygenase mechanism

commenced in the presence of both $^{17}O_2$ and $H_2^{18}O$. This was followed by the GC–MS analysis of retinol products which revealed equivalent quantities of oxygen derived from O_2 and H_2O (Leuenberger et al. 2001), and, therefore, BCO1 is also called as BCMO1 (β,β-carotene 15,15′-monooxygenase), the latter name now being preferably used. However, in the case of the dioxygenase reaction mechanism, two molecules of oxygen attack the double bond in the carotenoid substrate and form a dioxetane intermediate which is an unstable species. This dioxetane intermediate decays rapidly into two aldehyde products, and, therefore, water molecules are not required for the reaction. This mechanism was supported by isotope labeling experiment carried on CCD1 of *Arabidopsis thaliana* as discussed above. Further, in another study, a pure computational approach was used taking benefit of the available CCO structure. This computational approach indicated the dioxygenase mechanism to be preferable for the ACO enzyme. The dioxygenase mechanism was strengthened by the studies that suggested that the apparent 50 % incorporation of $^{17}O_2$ in previous studies was due to the exchange of isotopically labeled oxygen with bulk water.

2.5 Role and Biotechnological Applications of CCDs

2.5.1 Biotic and Abiotic Stress

Plants emit diverse volatiles in response to herbivore attack which can directly intoxicate, repel, or deter herbivorous insects or may alternatively protect the plant by attracting natural predators and parasitoids of these herbivores (Dicke and Loon 2000; Kessler and Baldwin 2001; Vancanneyt et al. 2001; Khan et al. 2000). Plant volatiles have already been proven to be a valuable tool for improving plant defense mechanisms (Khan et al. 2000), and volatiles produced by the activity of carotenoid cleavage can be utilized in improving plant defense. There are a number of reports where carotenoid cleavage products participate in various facets of plant defense mechanisms. In many plants like *Nypa fruticans* and *Lycopersicon esculentum*, CCDs control the emission of β- and α-ionone (Simkin et al. 2004 51 Azuma et al. 2002). While as β-ionone specifically repels beetles (*Anomala transvaalensis*), α-ionone is more active against *Macrodactylus subspinosus*, which can cause serious damage to the plants (Donaldson et al. 1990; Flath et al. 1994; McQuate and Peck 2001). *Halotydeus destructor* Tucker (Acari: Penthaleidae) fed less on *Trifolium glanduliferum* extracts that had high levels of β-ionones and other terpenes. A groundbreaking study, conducted by Wei et al. (2011), showed that transgenic *Arabidopsis* overexpressing CCD1 exhibited enhanced levels of α-ionone, a volatile apocarotenoid produced as a result of CCD1 dioxygenase activity. Moreover, when transgenic plants with elevated levels of α-ionone were tested for their interaction with *Phyllotreta cruciferae* (crucifer flea beetles), feeding damage was reduced in the transgenic plants compared to wild-type plants. This suggested that the volatile apocarotenoids deterred the insects from feeding on these plants. These compounds also protect the male gametophyte from pollen-consuming organisms which do not provide any benefit to plant fitness. Such volatile molecules can be produced in many plants through the introduction of CCDs from other plant sources by combinatorial biosynthesis approach. Since, CCDs show substrate promiscuity; they can cleave the same or the related substrates when introduced in other plants leading to the production of same apocarotenoids as that of the parent plant or novel and unnatural ones which may prove efficient in improving plant defense mechanisms. In order to adjust their membrane potential, some freshwater algae and cyanobacteria emit volatile compounds like β-cyclocitral and β-ionone under eutrophicated conditions or in response to photooxidative stress (Walsh et al. 1998). Some carotenoid cleavage products provide plants with competing advantage over surrounding plants like the presence of bioactive CCPs in sunflower or dihydroactinidiolide discharged around the rhizome of spikerush which inhibits the growth of surrounding plants. Secretion of α-ionone derivatives by maize roots inhibits the growth of the *Fusarium oxysporum*, and infection of tobacco by *Peronospora tabacina* leads to 50–600-fold increase in β-ionone which ultimately inhibits the sporulation of the fungus (Salt et al. 1986). It has been reported that CCD1 apocarotenoid products have antimicrobial activities (Fester et al. 1999). Therefore, expression of *CCD1* in all tissues may be important for plant defenses (Auldridge et al. 2006a). The role of CCPs in the colonization of plant roots by mycorrhizal fungus is supported by the accumulation of blumenin and a yellow pigment termed mycorradicin. Strigolactones, first discovered in root exudates of root-parasitic plant *Striga hermonthica*, produced under the nutrient-limiting conditions in roots, promote the growth of lateral roots and root hairs and inhibit the growth of lateral branches and buds in shoot. This strategy increases the nutrient uptake by the roots while as at the same time reduces the shoots' demand for the resources. Strigolactones have an effect, exclusively or in blend with other phytohormones, in the morphology of potato plants. Besides it also controls stolon development and maintenance of tuber dormancy. This is supported by the studies carried on CCD8-RNAi potato plants which have considerably more lateral and main branches than control plants. These plants also exhibit reduced stolon formation, have a dwarf phenotype, and lack flowering in case of the most severely affected lines (Pasare et al. 1999). These studies provide enough evidences that carotenoid cleavage dioxygenases can be utilized to improve the plant defense against both biotic and abiotic stresses. Suppression of some carotenoid cleavage dioxygenase such as CCD7 and CCD8 may give rise to a branched phenotype or alternatively the overexpression of some CCDs such as CCD4 may increase the production of volatile apocarotenoids like β-ionone. Both of these traits are biotechnologically feasible and agronomically valuable. In recent times, ABA accumulation and the expression of putative CCD genes were examined in mandarin plants. The results indicated that the transcription of CCD genes correlated with ABA accumulation pattern during severe water deficit conditions. The increasing knowledge about carotenoid cleavage enzymes will also enable ABA levels to be modified for improved drought and stress tolerance (Thompson et al. 2000). These studies have demonstrated the potential of engineering carotenoid degradation pathway and will open up opportunities for enhancing plant defense through biotechnological interventions.

2.5.2   Color, Flavor, and Aroma Compounds

Carotenoids and their cleavage products are involved in the development of aroma and flower color and can be undoubtedly exploited by metabolic engineering. The color of the flower has been modified in many plants including petunia (Van Der Krol et al. 1990), rose (Gutterson 1995), gerbera (Elomaa et al. 1993), carnation (Zuker et al. 2001), and chrysanthemum (Courtney-Gutterson et al. 1994). It is reported that several floricultural varieties have lost their fragrance by the traditional breeding programs because of a negative correlation between flower longevity and fragrance, and modification of floral scent and the color can prove highly beneficial for the floriculture industry (Vainstien et al. 2001; Gutterson 1993). Carotenoid cleavage dioxygenases are involved in the synthesis of a wide range of apocarotenoid molecules that are involved in imparting aroma, flavor, and color to flowers and fruits of a number of plants. The CCD enzymes are universal in plants and have been identified in many plants including strawberry, crocus (Rubio et al. 2008), coffee (Simkin et al. 2008), mandarin (Kato et al. 2006), melon (Ibdah et al. 2006), and grape (Mathieu et al. 2005) and several other species. Flowers of *Boronia megastigma* are valued for their orange-red flowers and appreciable floral oil, "Boronia Absolute." Acetone extracts of open flowers of the plant have been shown to contain several C_{27} derivatives (MacTavish et al. 2000). In *Crocus sativus*, the apocarotenoids picrocrocin, crocin, and safranal are responsible for the flavor, color, and aroma of the flower, respectively (Rubio et al. 2008; Bouvier et al. 2003; Simkin et al. 2004; Floss et al. 2008). The fruit color of *Bixa orellana*, color of *Citrus* fruit, and the variability in the color of *Lilium* sepals are due to the activity of carotenoid cleavage dioxygenases (Bouvier et al. 2003). In chrysanthemum, the suppression of *CmCCD4a* expression by RNAi turns the white petal color to yellow, which may be attributed to the degradation of carotenoids by CmCCD4a. Analysis of carotenogenic gene expression revealed that, all the genes encoding carotenoid biosynthesis enzymes are expressed in white chrysanthemum petals, indicating that carotenoids are biosynthesized in white petals (Ohmiya et al. 2006). The cyclic apocarotenoids β-ionone and β-damascenone and linear apocarotenoids including 6-methyl-5-hepten-2-one and geranylacetone are characterized as fruit/floral. Humans are intricately sensitive to these molecules. The cyclic apocarotenoids have very low odor thresholds than the linear. Thus, they are very important for the flavor of fruits and flowers. Except for β-damascenone, all of these apocarotenoids can be directly formed from their carotenoid precursor substrates in tomato by the action of two CCDs, which include LeCCD1A and LeCCD1B (Simkin et al. 2004). These enzymes are promiscuous in nature as far as their substrate specificity is concerned and recognize any carotenoid from ζ-carotene onward. It has been observed that the transgenic lines with knockdown of LeCCD1A and LeCCD1B have considerably reduced concentrations of all apocarotenoid volatiles, but they are not completely excluded (Simkin et al. 2004), suggestive of involvement of more than one additional CCDs in vivo.

Carotenoid cleavage dioxygenase play an important role in carotenoid turnover. Downregulation or inactivation of these enzymes can help to increase the carotenoid content of some important crop plants. The carotenoid content in *Solanum tuberosum* is completely dependent on the expression of carotenoid cleavage dioxygenase 4. Similarly, CCD4 is the major determinant in the accumulation of carotenoids and carotenoid-derived volatiles in peach fruit flesh (Brandi et al. 2011). Recent studies have shown that inactivation of CCD4 leads to the enhanced accumulation of carotenoids in *Prunus persica* (Ma et al. 2014). Carotenoid cleavage dioxygenases can be manipulated through various biotechnological strategies to improve the quality of agronomically important plants. Suppression or inactivation of CCDs on one hand may enhance the accumulation of carotenoids in many food plants, and over-expression of these enzymes in desired plants may lead to the accumulation of aroma, color, or flavor compounds. Combinatorial biosynthesis of apocarotenoids can also prove beneficial in the synthesis of desired apocarotenoids in transgenic plants wherein a CCD gene from one plant may be transferred to the other.

2.5.3   Pollination

Pollination is a critical step in the sexual reproduction of plants and is a major cost factor for the plants in which pollination is carried out by artificial techniques. Insufficient pollination can cause tremendous loss in the plants in which most of the ovules need to be fertilized to produce fruit of desired shape and size (Dudareva and Negre 2005). The advances in the identification and characterization of genes involved in the formation of volatile compounds in plants have formed the basis for biotechnological manipulation of plant volatiles. A number of attempts have been made to modify the floral scent to investigate the effect on insect behavior (Aharoni et al. 2003). Plants emit a number of volatile organic compounds to attract pollinators. During the course of evolution, plants have developed mutualistic relations with the pollinators wherein the pollinators get rewards for the act of pollination they are involved in. Pollination is valuable, for direct production of human-utilized plant products and for the successful reproduction of a plant as well (Dudareva and Negre 2005). CCDs are generally involved in the synthesis of volatile compounds that can attract the pollinators such as α- and β-ionone, safranal, pseudoionone, grasshopper ketone, etc. (Bouvier et al. 2005a, b). Expression of CCDs in transgenic plants can play an important role in the act of pollination by the production of pollinator attractants. In some cases, this approach may have the limitations as the production of a volatile in a transgenic plant may attract the enemies of the plant or even the enemies of the pollinators (Horiuchi et al. 2003). So, the introduction of CCD in any transgenic plant requires prior knowledge about the behavior of the natural predators of the plant. However, this approach may prove highly beneficial in some plants such as *Arnebia benthamii*, an important medicinal plant of Kashmir Himalayas. The plant produces seeds in its natural habitat, but when grown under nonnative habitat for commercial purposes, it does not produce seeds due to the unavailability of its pollinators. There are many other economically important plants that are not able to produce seeds due to the unavailability of pollinators when grown in nonnative habitats. Similarly, vanilla and coca are dependent on a few species of pollinators only and cannot be grown outside their natural habitats without additional expenses on artificial pollination techniques. Some plants do not produce pollinator attractants at all, while some of the plants produce them in their natural habitats only. Therefore, overexpression of CCDs in such plants may be utilized to produce substantial amounts of attractants for pollinators.

2.6 Conclusion

Carotenoid cleavage dioxygenases are involved in the production of a variety of bioactive molecules which play vital roles in plants. The overall structure of CCDs is conserved across plant kingdom. However, further studies are required to elucidate the complete mechanism of reaction. Owing to the rapid advances in genomics and metabolomics technologies, the identification of CCDs involved in synthesis of flavor volatiles and color compounds has speeded up in the past few years. Although our knowledge of regulation of the pathways is still rudimentary, transgenes capable of altering aroma, color, and carotenoid content are already available. It is likely that fruits and flowers with improved flavor and color will require coordinated regulation of multiple biosynthetic pathways. The discovery of carotenoid cleavage dioxygenases and their roles in some plants have great implications in other plants to improve their defense, flavor of their fruits, and the color and scent of the flowers. As CCDs have a significant impact on the carotenoid turn over, they can be manipulated to alter the carotenoid content of some important food plants. The growing evidences about the health benefits of apocarotenoids demand their production in higher quantities, and for that biotechnological strategies can prove very beneficial.

References

Aharoni A, Giri AP, Deuerlein S, Griepink F, de Kogel WJ, Verstappen FW, Bouwmeester HJ (2003) Terpenoid metabolism in wild-type and transgenic Arabidopsis plants. Plant Cell 15(12):2866–2884

Ashraf N, Jain D, Vishwakarma RA (2015) Identification, cloning and characterization of an ultra-petala transcription factor CsULT1 from *Crocus*: a novel regulator of apocarotenoid biosynthesis. BMC Plant Biol 15(1):25

Auldridge ME, Block A, Vogel JT, Dabney-Smith C, Mila I, Bouzayen M, Klee HJ (2006a) Characterization of three members of the Arabidopsis carotenoid cleavage dioxygenase family demonstrates the divergent roles of this multifunctional enzyme family. Plant J 45(6):982–993

Auldridge ME, McCarty DR, Klee HJ (2006b) Plant carotenoid cleavage oxygenases and their apocarotenoid products. Curr Opin Plant Biol 9(3):315–321

Azuma H, Toyota M, Asakawa Y, Takaso T, Tobe H (2002) Floral scent chemistry of mangrove plants. J Plant Res 115(1):0047–0053

Baba SA, Malik AH, Wani ZA, Mohiuddin T, Shah Z, Abbas N, Ashraf N (2015a) Phytochemical analysis and antioxidant activity of different tissue types of *Crocus sativus* and oxidative stress alleviating potential of saffron extract in plants, bacteria, and yeast. S Afr J Bot 31(99):80–87

Baba SA, Jain D, Abbas N, Ashraf N (2015b) Overexpression of Crocus carotenoid cleavage dioxygenase, CsCCD4b, in Arabidopsis imparts tolerance to dehydration, salt and oxidative stresses by modulating ROS machinery. J plant physiol 189:114–125

Baba SA, Mohiuddin T, Basu S, Swarnkar MK, Malik AH, Wani ZA, Abbas N, Singh AK, Ashraf N (2015c) Comprehensive transcriptome analysis of *Crocus sativus* for discovery and expression of genes involved in apocarotenoid biosynthesis. BMC Genomics 16(1):1

Bouvier F, Suire C, Mutterer J, Camara B (2003) Oxidative remodeling of chromoplast carotenoids identification of the carotenoid dioxygenase CsCCD and CsZCD genes involved in Crocus secondary metabolite biogenesis. Plant Cell 15(1):47–62

Bouvier F, Isner JC, Dogbo O, Camara B (2005a) Oxidative tailoring of carotenoids: a prospect towards novel functions in plants. Trends Plant Sci 10(4):187–194

Bouvier F, Rahier A, Camara B (2005b) Biogenesis, molecular regulation and function of plant isoprenoids. Prog Lipid Res 44(6):357–429

Brandi F, Bar E, Mourgues F, Horváth G, Turcsi E, Giuliano G, … Rosati C (2011) Study of 'Redhaven' peach and its white-fleshed mutant suggests a key role of CCD4 carotenoid dioxygenase in carotenoid and norisoprenoid volatile metabolism. BMC plant biol 11(1):1

Bugg TD, Ahmad M, Hardiman EM, Rahmanpour R (2011) Pathways for degradation of lignin in bacteria and fungi. Nat Prod Rep 28(12):1883–1896

Courtney-Gutterson N, Napoli C, Lemieux C, Morgan A, Firoozabady E, Robinson KE (1994) Modification of flower color in florist's chrysanthemum: production of a white flowering variety through molecular genetics. Nat Biotechnol 12(3):268–271

Dicke M, Loon JJ (2000) Multitrophic effects of herbivore-induced plant volatiles in an evolutionary context. Entomologia Experimentalis et Applicata 97(3):237–249

Donaldson JM, McGovern TP, Ladd TL (1990) Floral attractants for Cetoniinae and rutelinae (Coleoptera: Scarabaeidae). J Econ Entomol 83(4):1298–1305

Dudareva N, Negre F (2005) Practical applications of research into the regulation of plant volatile emission. Curr Opin Plant Biol 8(1):113–118

Elomaa P, Honkanen J, Puska R, Seppanen P, Helariutta Y, Mehto M, … Teeri TH (1993) Agrobacterium-mediated transfer of antisense chalcone synthase cDNA to Gerbera hybrida inhibits flower pigmentation. Bio/technology (USA)

Fester T, Maier W, Strack D (1999) Accumulation of secondary compounds in barley and wheat roots in response to inoculation with an arbuscular mycorrhizal fungus and co-inoculation with rhizosphere bacteria. Mycorrhiza 8(5):241–246

Flath RA, Cunningham RT, Liquido NJ, McGovern TP (1994) Alpha-ionol as attractant for trapping Bactrocera latifrons (Diptera: Tephritidae). J Econ Entomol 87(6):1470–1476

Floss DS, Schliemann W, Schmidt J, Strack D, Walter MH (2008) RNA interference-mediated repression of MtCCD1 in mycorrhizal roots of Medicago truncatula causes accumulation of C27 apocarotenoids, shedding light on the functional role of CCD1. Plant Physiol 148(3):1267–1282

Frusciante S, Diretto G, Bruno M, Ferrante P, Pietrella M, Prado-Cabrero A, Rubio-Moraga A, Beyer P, Gomez-Gomez L, Al-Babili S, Giuliano G (2014) Novel carotenoid cleavage dioxygenase catalyzes the first dedicated step in saffron crocin biosynthesis. Proc Natl Acad Sci 111(33):12246–12251

Gutterson NC (1993) Molecular breeding for color, flavor and fragrance. Sci Hortic 55:141–160

Gutterson N (1995) Anthocyanin biosynthetic genes and their application to flower color modification through sense suppression. Hortscience 30:964–966

Horiuchi JI, Arimura GI, Ozawa R, Shimoda T, Dicke M, Takabayashi J, Nishioka T (2003) Lima bean leaves exposed to herbivore-induced conspecific plant volatiles attract herbivores in addition to carnivores. Appl Entomol Zool 38(3):365–368

Huang FC, Horváth G, Molnár P, Turcsi E, Deli J, Schrader J, … Schwab W (2009) Substrate promiscuity of RdCCD1, a carotenoid cleavage oxygenase from Rosa damascena. Phytochemistry 70(4):457–464

Ibdah M, Azulay Y, Portnoy V, Wasserman B, Bar E, Meir A, … Tadmor Y (2006) Functional characterization of CmCCD1, a carotenoid cleavage dioxygenase from melon. Phytochemistry 67(15):1579–1589

Ilg A, Yu Q, Schaub P, Beyer P, Al-Babili S (2010) Overexpression of the rice carotenoid cleavage dioxygenase 1 gene in Golden Rice endosperm suggests apocarotenoids as substrates in planta. Planta 232(3):691–699

Kamoda S, Saburi Y (1993) Cloning, expression, and sequence analysis of a lignostilbene-α, β-dioxygenase gene from Pseudomonas paucimobilis TMY1009. Biosci Biotechnol Biochem 57(6):926–930

Kato M, Matsumoto H, Ikoma Y, Okuda H, Yano M (2006) The role of carotenoid cleavage dioxygenases in the regulation of carotenoid profiles during maturation in citrus fruit. J Exp Bot 57(10):2153–2164

Kessler A, Baldwin IT (2001) Defensive function of herbivore-induced plant volatile emissions in nature. Science 291:2142–2143

Khan ZR, Pickett JA, Berg JVD, Wadhams LJ, Woodcock CM (2000) Exploiting chemical ecology and species diversity: stem borer and striga control for maize and sorghum in Africa. Pest Manag Sci 56(11):957–962

Leuenberger MG, Engeloch-Jarret C, Woggon WD (2001) The reaction mechanism of the enzyme-catalyzed central cleavage of β-carotene to retinal. Angew Chem Int Ed 40(14):2613–2617

Ma J, Li J, Zhao J, Zhou H, Ren F, Wang L, Han Y (2014) Inactivation of a gene encoding carotenoid cleavage dioxygenase (CCD4) leads to carotenoid-based yellow coloration of fruit flesh and leaf midvein in peach. Plant Mol Biol Report 32(1):246–257

Mathieu S, Terrier N, Bigey F, Günata Z (2005) A carotenoid cleavage dioxygenase from Vitis vinifera L.: functional characterization and expression during grape berry development in relation to C13-norisoprenoid accumulation. J Exp Bot 56(420):2721–2731

McQuate GT, Peck SL (2001) Enhancement of attraction of alpha-ionol to male Bactrocera latifrons (Diptera: Tephritidae) by addition of a synergist, cade oil. J Econ Entomol 94(1):39–46

Moiseyev G, Takahashi Y, Chen Y, Gentleman S, Redmond TM, Crouch RK, Ma JX (2006) RPE65 is an iron (II)-dependent isomerohydrolase in the retinoid visual cycle. J Biol Chem 281(5):2835–2840

Oberhauser V, Voolstra O, Bangert A, von Lintig J, Vogt K (2008) NinaB combines carotenoid oxygenase and retinoid isomerase activity in a single polypeptide. Proc Natl Acad Sci 105(48):19000–19005

Ohmiya A, Kishimoto S, Aida R, Yoshioka S, Sumitomo K (2006) Carotenoid cleavage dioxygenase (CmCCD4a) contributes to white color formation in chrysanthemum petals. Plant Physiol 142(3):1193–1201

Pan Z, Zeng Y, An J, Ye J, Xu Q, Deng X (2012) An integrative analysis of transcriptome and proteome provides new insights into carotenoid biosynthesis and regulation in sweet orange fruits. J Proteome 75(9):2670–2684

Pasare SA, Ducreux LJ, Morris WL, Campbell R, Sharma SK, Roumeliotis E, … Qin X, Zeevaart JA (1999) The 9-cis-epoxycarotenoid cleavage reaction is the key regulatory step of abscisic acid biosynthesis in water-stressed bean. Proc Natl Acad Sci 96(26):15354–15361

Rubio A, Rambla JL, Santaella M, Gómez MD, Orzaez D, Granell A, Gómez-Gómez L (2008) Cytosolic and plastoglobule-targeted carotenoid dioxygenases from Crocus sativus are both involved in β-ionone release. J Biol Chem 283(36):24816–24825

Salt SD, Tuzun S, Kuć J (1986) Effects of β-ionone and abscisic acid on the growth of tobacco and resistance to blue mold. Mimicry of effects of stem infection by Peronosporatabacina Adam. Physiol Mol Plant Pathol 28(2):287–297

Schmidt H, Kurtzer R, Eisenreich W, Schwab W (2006) The carotenase AtCCD1 from Arabidopsis thaliana is a dioxygenase. J Biol Chem 281(15):9845–9851

Schwartz SH, Tan BC, Gage DA, Zeevaart JA, McCarty DR (1997) Specific oxidative cleavage of carotenoids by VP14 of maize. Science 276(5320):1872–1874

Simkin AJ, Schwartz SH, Auldridge M, Taylor MG, Klee HJ (2004) The tomato carotenoid cleavage dioxygenase 1 genes contribute to the formation of the flavor volatiles β-ionone, pseudoionone, and geranylacetone. Plant J 40(6):882–892

Simkin AJ, Moreau H, Kuntz M, Pagny G, Lin C, Tanksley S, McCarthy J (2008) An investigation of carotenoid biosynthesis in Coffeacanephora and Coffeaarabica. J Plant Physiol 165(10):1087–1106

Tan BC, Schwartz SH, Zeevaart JA, McCarty DR (1997) Genetic control of abscisic acid biosynthesis in maize. Proc Natl Acad Sci 94(22):12235–12240

Tan BC, Joseph LM, Deng WT, Liu L, Li QB, Cline K, McCarty DR (2003) Molecular characterization of the Arabidopsis 9-cis epoxycarotenoid dioxygenase gene family. Plant J 35(1):44–56

Thompson AJ, Jackson AC, Symonds RC, Mulholland BJ, Dadswell AR, Blake PS, … Taylor IB (2000) Ectopic expression of a tomato 9-cis-epoxycarotenoid dioxygenase gene causes overproduction of abscisic acid. Plant J 23(3):363–374

Van Der Krol AR, Mur LA, de Lange P, Mol JN, Stuitje AR (1990) Inhibition of flower pigmentation by antisense CHS genes: promoter and minimal sequence requirements for the antisense effect. Plant Mol Biol 14(4):457–466

Vancanneyt G, Sanz C, Farmaki T, Paneque M, Ortego F, Castañera P, Sánchez-Serrano JJ (2001) Hydroperoxide lyase depletion in transgenic potato plants leads to an increase in aphid performance. Proc Natl Acad Sci 98(14):8139–8144

Vogel JT, Walter MH, Giavalisco P, Lytovchenko A, Kohlen W, Charnikhova T, … Fernie AR (2010) SlCCD7 controls strigolactone biosynthesis, shoot branching and mycorrhiza-induced apocarotenoid formation in tomato. Plant J 61(2):300–311

Walsh K, Jones GJ, Dunstan RH (1998) Effect of high irradiance and iron on volatile odour compounds in the cyanobacterium Microcystis aeruginosa. Phytochemistry 49(5):1227–1239

Walter MH, Strack D (2011) Carotenoids and their cleavage products: biosynthesis and functions. Nat Prod Rep 28:663–692

Walter MH, Fester T, Strack D (2000) Arbuscular mycorrhizal fungi induce the non-mevalonate methylerythritol phosphate pathway of isoprenoid biosynthesis correlated with accumulation of the 'yellow pigment'and other apocarotenoids. Plant J 21(6):571–578

Westerkamp C, Gottsberger G (2000) Diversity pays in crop pollination. Crop Sci 40:1209–1222

Young PR, Lashbrooke JG, Alexandersson E, Jacobson D, Moser C, Velasco R, Vivier MA (2012) The genes and enzymes of the carotenoid metabolic pathway in Vitis vinifera L. BMC Genomics 13(1):1

Zeevaart JA, Heath TG, Gage DA (1989) Evidence for a universal pathway of abscisic acid biosynthesis in higher plants from 18O incorporation patterns. Plant Physiol 91(4):1594–1601

Zhang M, Leng P, Zhang G, Li X (2009) Cloning and functional analysis of 9-cis-epoxycarotenoid dioxygenase (NCED) genes encoding a key enzyme during abscisic acid biosynthesis from peach and grape fruits. J Plant Physiol 166(12):1241–1252

Zuker A, Tzfira T, Scovel G, Ovadis M, Shklarman E, Itzhaki H, Vainstein A (2001) RolC-transgenic carnation with improved horticultural traits: quantitative and qualitative analyses of greenhouse-grown plants. J Am Soc Hortic Sci 126(1):13–18

Chapter 3
Pharmacological Importance of *Crocus sativus* Apocarotenoids

Abstract *Crocus sativus* L. is an important medicinal plant belonging to family Iridaceae. It has been used in traditional medicine since time immemorial. The traditional medicinal uses have also validated scientifically. Different bioactivities like anticancer, neuroprotective, anti-inflammatory, and cardioprotective have been ascribed to *C. sativus*. These properties are speculated to be due to the presence of several carotenoids and their apocarotenoid products. This chapter provides an in-depth knowledge about the traditional medicinal uses of saffron and their subsequent pharmacological validation.

Keywords Apocarotenoids • Anticancer • Neuroprotective • Hepatoprotective

3.1 Introduction

Natural products have been accepted as central source of pharmacologically effective medicines and are in the cards to continue as sustainable sources of novel drug leads. Approximately 60% of the world's population is almost exclusively dependent on plants and plant-derived products (Farnsworth 1994). Utilizing the leaves, flowers, stems, berries, and roots of plants is not uncommon in different traditional systems of medicine across the globe and has proved beneficial in the prevention of various diseases and disorders and overall functioning of a particular organ or whole body.

   C. sativus L. is an important medicinal plant and has been used widely in tropical and subtropical countries for treating a number of ailments. The use of saffron for medicinal purposes dates back to the Greek and Roman periods. It has been used to treat stomach disorders, flatulence, coughs, colic, insomnia, feminine disorder, and heart disease, as anodyne and tranquilizer, and for its emetic properties (Abdullaev et al. 2003). Saffron has been used in Chinese traditional medicine and is also mentioned in "Yi-Lin-Ji-Yao," a traditional Chinese medical book composed during the Ming Dynasty (sixteenth century). In this Chinese medical book, the effects of saffron on promotion of blood circulation to remove blood stasis have been extensively described (Abdullaev and Frenkel 1999). Similarly, in Islamic culture, one of the most important books of medicine is al-Qanun or Canon of Medicine, which was written by Avicenna. This book describes around 800 different natural drugs and contains a monograph dedicated to the medicinal properties of saffron (Hosseinzadeh

© Springer Science+Business Media Singapore 2016   39
S.A. Baba, N. Ashraf, *Apocarotenoids of Crocus sativus L: From biosynthesis to
pharmacology*, SpringerBriefs in Plant Science, DOI 10.1007/978-981-10-1899-2_3

Table 3.1 Uses of saffron in traditional medicine

Systems and effects	Traditional uses
Analgesic and anti-inflammatory	Earache, toothache, swelling, otitis, anal pain, gout, cancer pain, gingivitis, discomfort of teething in infants
Cardiovascular system	Cardiac stimulant, removes blockages of vascular
Central nervous system	Narcotic, antihysteric, CNS stimulant, hypnotic, mental disease, neurasthenia sedative, anticonvulsant
Eye disease	Painful eye, lacrimation, day blindness, corneal disease and cataract, purulent eye infection, pterygium, poor vision
Gastrointestinal	Stomachic, decreased appetite, treatment of hemorrhoid
Genitourinay system	Abortive, treatment of amenorrhea, aphrodisiac, impotency
Infection disease	Antibacterial, antiseptic, antifungal, measles, smallpox
Respiratory system	Asthma, bronchitis, expectorant, pertussis, dyspnea

and Nassiri-Asl 2013). In Indian Ayurvedic medicine saffron is recognized as an adaptogen (Kianbakht and Mozaffari 2009). Saffron syrup, saffron glycerin, and saffron extract have been mentioned in English pharmaceutical codex. Saffron is known to facilitate digestion and its essential oil is considered useful in insomnia. In South Asia saffron is widely used for liver, kidney, and vesica disease and for treatment of cholera. For the treatment of dermal diseases such as impetigo, saffron tincture is considered useful. Some of the important uses of saffron in traditional medicine are listed in Table 3.1.

Traditional knowledge of medicinal properties of saffron has attracted scientists across the globe, and during the last few decades, several molecules with tremendous pharmacological importance have been isolated. These bioactive molecules include the unique apocarotenoids of saffron such as crocetin, crocin, picrocrocin, and safranal. These molecules have been shown to possess a diversity of pharmacological activities ranging from neuroprotective to anticancer activities (Baba et al. 2015).

3.2 Neuroprotective Activity

Neurodegeneration is a common consequence of several nervous system diseases. These diseases are disturbing and costly to treat. The current annual costs for their treatment exceed several hundred billion dollars in the United States alone. Besides, the existing treatments for these diseases are inadequate. The incidence of age-related neurodegenerative diseases is increasing rapidly as population demographics change adding further to the urgency of the problem.

C. sativus apocarotenoids have displayed many pharmacological effects on the nervous system. These pharmacological activities include anxiolytic properties, antidepressant effects, aphrodisiac activity, improvement of learning and memory, and lessening of physical signs of morphine withdrawal (Alavizadeh and Hosseinzadeh 2014) (Tables 3.2, 3.3 and 3.4). Neurodegenerative disorders are often associated with memory and learning impairments. Studies have shown that

Table 3.2 List of some recently carried in vitro studies on *C. sativus* apocarotenoids

Apocarotenoid	Effect	Cell line	Probable mechanism	References
1. Crocin	Anticancer	Adenocarcinoma gastric cancer cells (AGS)	Suppression of Bcl-2 and activation of Bax	
	Anti-inflammatory	Murine macrophage RAW264.7	Suppression of LPS-stimulated expression of nitric oxide synthase by inducing heme oxygenase-1 expression	
	Neuroprotective	PC-12 cell line	Inhibition of intracellular ROS	Mehri et al. (2012)
2. Crocetin	Anticancer	Human cervical cancer HeLa cell line	Inhibition of COX-2 expression	Chen et al. (2015)
	Anticancer	Esophageal squamous cell carcinoma KYSE-150	Increased expression of proapoptotic Bax and activated caspase 3	
	Neuroprotective	Hippocampal HT22 cell line	Impairing the antioxidant defense and detoxification systems	
3. Safranal	Cardioprotective	H9c2 cardiomyocytes	Reduction of oxidative stress	
	Antitumor	K-562 human chronic myelogenous leukemia (CML) cells	Inhibition of Bcr-Abl protein	Geromichalos et al. (2012)

crocin (10–30 μM) has the potential to provoke the inhibition of long-term potentiation tempted by ethanol in hippocampal neurons. Crocin provokes the suppression by acting on N-methyl-D-aspartate receptors. In an in vivo study, it was observed that crocin repressed hyoscine-induced learning discrepancies and decreased performance abilities at both low and high doses (50–200 mg/kg) (Hosseinzadeh and Ziaei 2006). This was further confirmed by the involvement of crocin in modulating storage and/or retrieval of information as reported by Pitsikas et al. (2007). Safranal has also been shown to have significant effects on memory like that of crocin. Hosseinzadeh and Ziaei (2006) also reported that safranal

Table 3.3 Some recent *in vivo* studies carried out on *C. sativus* apocarotenoids

Apocarotenoid	Effect	Dosage (mg/kg)/Animal/Route	Probable mechanism	References
1. Crocin	Learning and memory	5–25, Wistar rats ($n = 7$), i.p.	Antioxidant effects	Hosseinzadeh et al. (2012)
	Morphine withdrawal	400–600, NMRI mice ($n = 7$), i.p.		Imenshahidi et al. (2011)
	Antidepressant	12.5–50, Wistar rats ($n = 30$), i.p.	Increase in CREB, BDNF, and VGF	Hassani et al. (2014)
	Prevention of obsessive–compulsive disorder	30–50, Wistar rats ($n = 8$), i.p.	Interaction between crocins and the serotonergic system	
	Anticonvulsive	12.5–100 µg, Wistar rats ($n = 6$), i.c.v.	Participation of GABA–benzodiazepine receptor complex	Tamaddonfard et al. (2012)
	Protection of myocardial injury	10–40, male Sprague Dawley rats ($n = 10$), oral	Increased superoxide dismutase and catalase enzyme activities and decreased malondialdehyde (MDA)	Dianat et al. (2014)
	Antitumor effect	50–100, mice ($n = 6$), i.v.		Rastgoo et al. (2013)
	Hepatoprotective	200, Wistar rats ($n = 8$), i.p.	Reduction of $BeCl_2$-induced oxidative stress and mRNA expression of antioxidant genes	El-Beshbishy et al. (2012)
	Gastroprotective	50, rats ($n =$), i.p.	Increase in levels of prostaglandin E2 (PGE2) and interleukin-6 (IL-6) and decrease in ethanol-elevated tumor necrosis factor-alpha (TNF-α) level, myeloperoxidase activity, and heat shock protein 70 mRNA	

Apocarotenoid	Effect	Dosage (mg/kg)/Animal/Route	Probable mechanism	References
2. Crocetin	Anticancer	10–40, female Kunming strain mice ($n = 12$), oral	Increase in maleic dialdehyde, polymorphonuclear cells (PMN), interleukin-1β (IL-1β), and tumor necrosis factor-α (TNF-α)	Chen et al. (2015)
	Anticancer	50–100, Wistar rats ($n = 5$), i.p.	Suppression of Bcl-2 and activation of Bax	Bathaie et al. (2013)
3. Safranal	Antihypertensive	1–4, Wistar rats ($n = 42$), i.p.	Effect on mean systolic blood pressure (MSBP) and heart rate (HR)	
	Morphine withdrawal	0.0085–0.15 ml/kg ($n = 8$), i.p.	Opioid partial agonism	Hosseinzadeh and Jahanian (2010)
	Antioxidant	0.5, Wistar rats ($n = 5$), i.p.	Enhancement of antioxidant defenses	
	Prevention of retinitis pigmentosa	400, homozygous P23H line-3 albino rats, i.p.	Slowing of photoreceptor cell degeneration	

Table 3.4 List of some clinical studies carried out on *C. sativus*

Disease/disorder	Design of study	Dosage/duration	Agent	Reference
Depression	Double blind	30 mg/day, p.o.	*C. sativus* (petal)	Akhondzadeh et al. (2005)
		30 mg/day, p.o. 6 weeks	*C. sativus* (stigma)	
Erectile dysfunction	Open	200 mg/day, p.o. 10 days	*C. sativus* (stigma)	
Alzheimer's disease	Double blind	30 mg/day, p.o. 16 weeks	*C. sativus* (stigma)	Akhondzadeh et al. (2010a)
Alzheimer's disease	Double blind	30 mg/day, p.o.	*C. sativus* (stigma)	Akhondzadeh et al. (2010b)
		10 mg/day, p.o. 22 weeks	Donepezil	
Age-related macular degeneration		20 mg/day, p.o. 8 months	*C. sativus*(stigma)	
Alzheimer's disease	Double blind	30 mg/day, p.o.	*C. sativus* (stigma)	Farokhnia et al. (2014)
		20 mg/day, p.o. 16 weeks	Memantine	
Depression	Double blind	30 mg/day, p.o. 4 weeks	Crocin	Talaei et al. (2015)

(0.2 ml/kg) repaired the weakening activity of scopolamine (1 mg/kg) on memory in rats in the Morris water maze model. Besides it displayed no effect on intact memory. This was possibly because of its sedative effect through the BDZ site. Since ROS have shown to be involved in neurodegenerative diseases, antioxidants are believed to be promising neuroprotectors. This is due to their ability to scavenge ROS (Ochiai et al. 2004a, b; Saleem et al. 2006). A study has stated that treatment of PC-12 cells with crocin repressed cell membrane lipid peroxidation. Besides, it also restored intracellular superoxide dismutase (SOD) activity more effectively than α-tocopherol at same concentration (Ochiai et al. 2004a, b). This was found to be due to the potential of crocin to combat ischemic stress-induced neural cell death by enhancing GSH activities and precluding the initiation of c-Jun NH2-terminal kinase (JNK) pathway (Ochiai et al. 2004a, b). Anti-ischemia activity of crocin was also established by an in vivo investigation. In this study pretreatment of mice with crocin considerably repressed oxidative reactions. Further, crocin caused inhibition of matrix metalloproteinase-9 (MMP-9) expression in mice. This effect was observed within 20 min of bilateral common carotid artery occlusion proceeded by 24 h of reperfusion (Dong et al. 2009; Zheng et al. 2007). The anti-ischemic effects of safranal have also been reported. Administration of safranal (0.1–0.5 mL/kg) to rats intraperitoneally (i.p.) once daily for 14 days was followed by induction of ischemia on the 15th day by one-stage ligation of the left anterior descending coronary artery. Subsequently reperfusion was done for 60 min.

The results of this study revealed that safranal decreased infarct size and improved left ventricular operation and myocardium hemodynamic status. The results indicated that safranal enhanced phosphorylation of Akt/GSK-3b/eNOS and reduced IKK-b/NF-κB protein expressions proceeding ischemia–reperfusion in myocardium. Due to capacity of safranal to upregulate Bcl-2 transcript levels and down-regulate Bax and caspase 3 transcript levels, it has the ability to act as an efficient antiapoptotic molecule. In a model of ischemia–reperfusion injury, administration of safranal (0.1, 0.25, and 0.5 ml/kg, i.p.) before the injury considerably decreased thiobarbituric acid reactive species, and it was further established by histopathological investigations (Sadeghnia et al. 2008).

Depression is a serious and widespread mental disorder. It has been predicted to be one of the ten leading causes of disabilities that will affect up to 21 % of the world population by 2020 (Hassani et al. 2014). The antidepressant effects of *C. sativus* apocarotenoids were evaluated in acute preclinical studies and shown to be significantly more beneficial than placebo (Karimi et al. 2001; Hosseinzadeh et al. 2007; Wang et al. 2010). This activity was further examined in *C. sativus* stigma and petal aqueous and ethanolic extracts through forced swimming tests (FST) on mice (Karimi et al. 2001). It was observed that this effect was possibly facilitated by safranal (0.05–0.15 ml/kg) and crocin (50–600 mg/kg) by inhibition of dopamine, norepinephrine, and serotonin uptake. Afterward, it was revealed by behavioral models that antidepressant-like effect of *C. sativus* stigma aqueous extract may be due to the presence of crocins 1 and 2 (Wang et al. 2010). Besides being an effective antidepressant, some studies have shown that crocin may possess anxiolytic activity. In another in vivo study, administration of crocin (12.5, 25, and 50 mg/kg), imipramine (10 mg/kg; positive control), and saline (1 mL/kg; neutral control) intraperitoneally (i.p.) to male Wistar rats for 21 days was done. The antidepressant activity was then investigated by FST on day 21 after injection. The results revealed that crocin considerably decreased the immobility time in the FST. Western blots indicated that 25 and 50 mg/kg of crocin enhanced the levels of CREB and BDNF considerably and in a dose-dependent manner. All doses of crocin enhanced the VGF levels dose-dependently. Levels of p-CREB enhanced significantly by 50 mg/kg dose of crocin. Only 12.5 mg/kg crocin could considerably upregulate the expression levels of BDNF. However, no effects were reported in CREB and VGF expression levels in all of the groups (Hassani et al. 2014). Recently, in a randomized, double-blind, placebo-controlled, pilot clinical trial, addition of crocin tablets (30 mg/day, 15 mg BID) augmented the activity of selective serotonin reuptake inhibitor (SSRI) during the treatment of patients with mild to moderate depression. Moreover, because of the absence of substantial side effects, crocin was reported to be an effective therapeutic adjuvant (Talaei et al. 2015). It has been demonstrated that *Crocus* apocarotenoids like crocetin efficiently cross the blood-brain barrier and lessen the danger of neurodegenerative disorders. Co-incubation of hippocampal HT22 cell line with 5 μM crocetin and CuO nanoparticles significantly attenuated the CuO-induced toxicity.

C. sativus apocarotenoids have also been found to be useful in the treatment of Alzheimer's and Parkinson's diseases. The principal carotenoid cleavage product, *trans*-crocin 4 which is the digentobiosyl ester of crocetin, repressed A-beta fibril-logenesis produced by the oxidation of the amyloid beta peptide fibrils in Alzheimer's disease. Parkinson's disease is basically featured by the degeneration of neurons in the substantia nigra by ROS or by various chemicals like 6-hydroxy dopamine resulting in the death of neurons. In experimental rats which were pretreated with crocetin, an enhancement in the antioxidant potential of enzymes was observed. This was preceded by protection against the harmful effects of 6-hydroxy dopamine, sug-gesting that it may prove as an efficient treatment to overcome this distressing dis-ease (Xuan et al. 1999; Bhargava 2011). The above studies provide strong evidences that *Crocus* apocarotenoids prevent the onset or progression of nervous system-related disorders. However, further studies are required to find out the exact mecha-nism of action by which *C. sativus* apocarotenoids exert their beneficial effects.

3.3 Anticancer Activity

Cancer is one of the main health problems around the globe and is presently the second principal reason of death in the United States. Cancer is predicted to outdo heart diseases as the leading cause of death in the coming few years. In order to combat such devastating diseases, new drugs with better activity and/or less toxicity need to be developed, and for this, natural products are considered to be good can-didates. Recently, carotenoids have been shown to modulate expression of some key cancer-related genes. *C. sativus* apocarotenoids have tremendous anticancer and tumoricidal potential (Tables 3.2 and 3.3). It has been reported that treatment of HeLa cells with crocin resulted in decreased cytoplasm, cell shrinkage, and apopto-sis, while crocetin and safranal exhibited an insignificant part in the cytotoxic activ-ity of saffron extract (Escribano et al. 1996). A long-term treatment with crocin on colorectal cancer cells under in vitro conditions revealed the strong cytotoxic effect of crocin on human and animal adenocarcinoma cells which was also associated with morphological alterations in cells. Further, it was observed that degree of gly-cosylation has no effect on inhibition of growth of cancer cells (Aung et al. 2007; Chryssanthi et al. 2007). At molecular level, crocin has been found to induce changes in expression of genes involved in cell cycle including a decrease of Bcl-2 and increase of Bax expression. This leads to considerable increase in the number of cells at G0/G1 phase and ultimately increases the percentage of cell apoptosis (Lv et al. 2008; Xu et al. 2010; Zhao et al. 2008). This was also confirmed by another investigation wherein human pancreatic and tongue squamous cancer cell lines were administrated with 10 µg/L and 0.4 mM of crocin, respectively (Bakshi et al. 2010; Sun et al. 2011). In another study crocin encapsulated in liposomes exhibited higher IC_{50} values (IC_{50} values after 48 h: 0.61, 0.64, and 1.2 mM) than free crocin (IC_{50} after 48 h: 1.603 mM) in HeLa cells (Mousavi et al. 2011). Although there are numerous in vitro studies on the anticancer and tumoricidal effects of cro-cin, there are only a few reports of in vivo studies. Konoshima et al. (1998) reported that crocin delayed the formation of mouse skin papillomas under in vivo conditions

(Konoshima et al. 1998). Similarly, subcutaneous administration of crocin enhanced the life span of rats and reduced the growth of tumor strongly in female rats, and no major toxic effects were observed (Garc-Olmo et al. 1999). The antitumor effect of PEGylated nanoliposomes enclosing crocin (50 and 100 mg/kg, intravenously) was observed recently in BALB/c mice with C26 colon carcinoma. The results revealed that encapsulation of crocin in liposomes enhanced its antitumorigenic effect (Rastgoo et al. 2013). Similar to crocin, crocetin has also been reported to have anticancer and tumoricidal activities (Abdullaev and Espinosa-Aguirre 2004; Moghaddasi 2010). reported that treatment of the three different types of cancer cells (cervical cancer cell line HeLa, non-small cell lung cancer cell line A549, and ovarian cancer cell line SKOV3) with crocetin (60–240 µmol/L) for 48 h considerably reduced their proliferation concentration-dependently. Crocetin (240 µmol/L) brought about the cell cycle arrest via p53-dependent and p53-independent pathways accompanied with p21$^{WAF1/Cip1}$ induction. In another study, anticancer effects of crocetin on the esophageal squamous cell carcinoma cell line KYSE-150 showed that crocetin significantly inhibited the proliferation of the cells for 48 h in a concentration-dependent manner, and the inhibition of proliferation was accompanied by S phase arrest. Recently, for the first time, crocetin was reported to have anti-inflammatory effect by suppressing the level of IL-1 β and TNF- α as well as PMN activity in a methylcholanthrene (MCA)-induced uterine cervix tumorigenesis murine model system. Further, crocetin dose-dependently decreased COX-2 production in cervical cancer cells (Chen et al. 2015). Similarly, in an in vivo study, the therapeutic effect of crocetin on 1-methyl-3-nitro-1-nitrosoguanidine (MNNG)-induced gastric cancer in rats has also been investigated. Crocetin caused reduction in AGS cell proliferation in dose- and time-dependent manner which was reported to be due to the reduction of Bcl-2 and increase of Bax expression. However, these effects were not reported in normal human fibroblast (HFSF-PI3) cells (Bathaie et al. 2013). Taken together, apocarotenoids of *C. sativus* may serve as novel leads for development of anticancer molecules. However, further clinical studies are required to establish *Crocus* apocarotenoids as good candidates for drug development. Since development of new drugs is limited by development of drug resistance which is responsible for treatment failure in more than 90 % of patients with metastatic disease, endeavors toward the development of new drugs are very much required.

3.4 Cardioprotection

Cardiological disorders are currently considered as major causes of death, claiming around 17 million lives every year. Epidemiological investigations have shown that diets which are rich in antioxidants are associated with a low risk of cardiovascular disease. *C. sativus* apocarotenoids are considered as strong antioxidants, and their cardioprotective potential is well established in literature (Shiping et al. 1999; Lee et al. 2005; Sheng et al. 2006; Du et al. 2005; Goyal et al. 2010). Crocin has been reported to inhibit cholestane-3b,5a,6b-triol-induced apoptosis which was attributed to increase in Bcl-2/Bax expression ratio (Liu and Qian 2005). However, Xu et al. (2007) further confirmed the earlier observations and concluded that inhibition of apoptosis by crocin had a vital role in the reduction and regression of atherosclerosis. Earlier, a study had reported cytoprotective activity of crocin (1–10 μM) against H_2O_2-induced endothelial cell injury dose-dependently which certainly added to its use in cardiovascular-related disorders (Xu et al. 2009). Goyal et al. (2010) reported the cardioprotective potential of crocin in isoproterenol (ISO)-induced cardiotoxicity through modulation of oxidative stress (Goyal et al. 2010). This was further confirmed in another study wherein cardioprotective effect of crocin on antioxidant capacity was studied in comparison to vitamin E in ischemia–reperfusion of isolated rat hearts. Randomly selected seven groups of Sprague Dawley rats were orally administrated with different doses of crocin (10, 20, and 40 mg/kg) or vitamin E (100 mg/kg) and a combination of crocin (40 mg/kg) with vitamin E (100 mg/kg) for 21 days. The heart was quickly excised, transferred to a Langendorff apparatus at constant pressure, and subjected to 30 min of global ischemia followed by 60 min of reperfusion. It was observed that superoxide dismutase (SOD) and catalase (CAT) enzyme activities were enhanced and malondialdehyde (MDA) was reduced in animals treated with crocin (40 mg/kg) and vitamin E (100 mg/kg). Further, there was a considerable enhancement in postischemic recovery of antioxidant activity during reperfusion in rats that received a combination of crocin (40 mg/kg) and vitamin E (100 mg/kg) (Dianat et al. 2014).

Like crocin, crocetin also exhibited the ability to prevent the development of atherosclerosis in quail (He et al. 2007a, b). Crocetin was also reported to reduce the level of cardiac marker—lactate dehydrogenase activity—and enhance mitochondrion potential in a cardiac myocyte pretreated with noradrenaline. This study suggests the cardioprotective activity of crocetin (Abe et al. 1999). In yet another investigation, crocetin due to its efficient antioxidant capacity prohibited norepinephrine-induced cardiac hypertrophy. This effect is due to increase in the expression levels of the antioxidant enzymes which include myocardial SOD, glutathione peroxidase, CAT, and improved norepinephrine-induced myocardial pathological histological effects (Shen et al. 2004). In another in vivo study, crocetin was administrated to rabbits to evaluate its outcome on the progression of atherosclerosis. Three different diet sets for eight weeks were randomly allotted to New Zealand white rabbits. The diets included a standard diet, a high lipid diet, or a high lipid + crocetin diet. The high lipid diet group showed hypercholesterolemia and atherosclerosis. Nevertheless, the crocetin-supplemented group lessened the harmful health effects of a high lipid diet. It was observed that there was a considerable variation in the plasma lipid levels and high density lipoprotein cholesterol between the high lipid diet and crocetin groups. However, there was no visible reduction in the atheroma, aorta cholesterol deposits, foam cells, and atherosclerotic injuries in the crocetin-fed group. It was then proposed that nuclear factor kappa B (NF-kB) initiation in the aortas is prevented by antioxidants like crocetin, and this in turn reduces the vascular cell adhesion molecule-1 transcript levels (Zheng et al. 2005). In yet another study, crocetin was reported to provide protection against myocardial ischemia–reperfusion injury (MIRI) in rats by preventing ROS production, hampering inflammation, and inhibiting myocardium apoptosis. In a rat model, it was observed that pretreatment with crocetin (50 mg/kg) decreased the cardiac injury, oxidative stress, and inflammation in comparison with that of the non-treated rats, as depicted by the reduced levels of infarct size, creatine kinase-MB (CK-MB) expression, MDA, and tumor necrosis factor-alpha (TNF-α) activity and the enhanced levels of total SOD and inflammation cytokine interleukin-10 activity. Crocetin activation also reduced the TUNEL staining-positive percentage and Bax transcript levels and increased Bcl-2 and eNOS transcripts and NO production, thus suggesting that crocetin may reduce the apoptotic damage (Wang et al. 2014).

3.5   Hepatoprotective Activity

Saffron and its constituents have also been shown to have hepatoprotective activity. It has been reported that intraperitoneal injection (i.p.) of crocin (200 mg/kg) avoids iron-induced liver injury in rats. This was observed to be revealed by considerable effects in the liver functioning indices, hyperammonemia and reduced serum urea levels (EL-Maraghy et al. 2009). In another study, Sun et al. 2013 showed that crocin confers protection against cisplatin-induced hepatotoxicity by reducing oxidative stress and preventing the stimulation of tumor protein 53 (p53), phospho-p38-mitogen-activated protein kinase (MAPK), and cleaved caspase 3 (Sun et al. 2013). This was further confirmed by another study wherein crocin was found to ameliorate CCl4-induced liver injury via inhibition of inflammatory cytokines, caspase 3, and oxidative stress along with modulation of liver-metabolizing enzymes favoring elimination of CCl4 toxic metabolite (Bahashwan et al. 2015). Oral administration of crocin has also been shown to eliminate hepatic injury by scavenging ROS in streptozotocin (STZ)-induced diabetic rats and to decrease the blood glucose levels significantly by hypoglycemic activity (Altinoz et al. 2015)

3.6   Immunological Functions of Saffron Carotenoids and Apocarotenoids

Carotenoids have shown two positive effects to human health via augmentation of the immune response and by lessening the danger of degenerative diseases (Kadian and Garg 2012). Carotenoids modulate immune responses by increasing natural killer cell (NK cell) activity, enhancing the lymphocyte response to mitogens, preventing immune cells from reactive species produced by their own, and increasing total WBCs and CD4/CD8 ratio in HIV-infected humans (Jain et al. 2013). For instance, β-carotene increases the receptors on WBCs for major histocompatibility complex class I (MHC-I) (Namin et al. 2009). Several studies have indicated saffron's capacity to protect cells from ROS by scavenging them, which is mainly attributed to two main saffron apocarotenoids: crocin and crocetin (Al-Qudsi and Ayedh 2012). The activity of crocin in activation of dendritic cells that increase the proliferation of T cells in mononuclear cells isolated from bone marrow of children suffering from leukemia has been reported (Noureini and Wink 2012). Since cancer often grows in swollen tissues, inflammatory condition is considered to be closely associated with carcinogenesis. Examples include liver and esophageal cancer, chronic hepatitis and gastritis (*Helicobacter pylori*), and inflammatory bowel disease (Tanaka et al. 2012). Since, nuclear factor-kB (NF-kB) signaling pathway has a major role in inflammation-associated carcinogenesis, the molecules that suppress NF-kB transcription may prove useful for cancer chemoprevention (Kawabata et al. 2012). As it is well established, inflammatory genes such as COX-2, iNOS, tumor necrosis factor-α (TNF-α), IL-6, and IL-1β are the most prevalent target genes taking part in the stimulation of NF-kB which is accompanied with several chronic inflammatory diseases (Kawabata et al. 2012; Tanaka et al. 2012). Studies have reported that dietary intake of crocin inhibits colitis and colitis-related colon carcinogenesis in mice induced by certain chemicals, by downregulating the expression mRNA of several pro-inflammatory cytokines and inducible inflammatory enzymes (Kawabata et al. 2012). A recent study demonstrated that carotenoids like astaxanthin also suppress the expression of inflammatory cytokines, NF-kB, besides inhibiting inflammation-associated colon carcinogenesis in mice (Tanaka et al. 2012). Further it has been shown that a humoral immune response decreases circulating carotenoids and carotenoid-based coloration in case of i.p. injection of lipopolysaccharide of *E. coli* which stimulates an inflammatory response. This is preceded by antibody generation, reduced plasma carotenoids, and carotenoid-based coloration. These investigations therefore back the view that an immune response mounting leads to draining of carotenoids from the bloodstream (Perez-Rodriguez et al. 2008). In an independent study, transplantable mammary tumor model in BALB/c mice was used to study the mechanism of action of dietary lutein against tumor growth. It was observed that dietary intake of lutein upregulated IFN-γ mRNA transcripts but downregulated the transcripts of IL-10 in splenocytes of mice with tumor, providing a clue toward the inhibitory effect of lutein against tumor growth. IFN-γ generated by activated T cells and NK cells has several immunoregulatory actions. It is a strong activator of macrophages and MHC class II molecules. In contrast, IL-10 prevents IFN-γ generation, antigen presentation, and IL-1, IL-6, and TNF-α formation by macrophages. The antitumor and immunomodulatory effects of lutein provide a clue toward the participation of subcellular processes including apoptosis, angiogenesis, and gene regulation (Chew and Park 2004). Intake of lutein through diet has been observed to be accompanied by reduced apoptosis in blood leukocytes of tumor-bearing mice in comparison with control. This suggests an increased immune status. On the other hand, apoptosis in tumor cells was augmented due to the intake of lutein which is suggestive of increased death of tumor cells. Moreover, the results indicated a choosy action of lutein which may be due to decreased apoptosis in immune cells but enhanced apoptosis in tumor cells (Chew and Park 2004). Additionally, dietary intake of lutein enhanced the transcript levels of proapoptotic genes p53 and Bax but downregulated the transcription of the antiapoptotic gene

Bcl-2, leading to higher Bax/Bcl-2 ratio in tumors. Moreover, it has been reported that carotenoids can affect immune function by their capacity to control membrane fluidity and gap junction communication. Studies have provided some convincing evidences that the effect of carotenoids on immunity and other diseases may be attributed to their ability to scavenge reactive oxygen species. The effect of carotenoids on immune response is dependent not only on the types and concentrations of carotenoids but also on cell types and animal species involved in the experiment (Chew and Park 2004).

One of the theories of immune regulation involves homeostasis between T-helper 1 (Th1) and T-helper 2 (Th2) activity. It has been reported that various nutrients and hormones affect Th1/Th2 balance such as plant sterols/sterolins and the minerals (e.g., zinc) (Bani et al. 2011). The immunomodulatory activity of *C. sativus* was recently reported on Th1 and Th2 limbs of the immune system. It was observed that oral administration of alcoholic extract of *C. sativus* (ACS) leads to a significant increase of CD19+ B cells and IL-4 cytokine which is considered as a trademark cytokine of Th2 pathway. Further increased levels of IgG-1 and IgM antibodies were also reported. However, no significant effect was observed on the transcript levels of the Th1 cytokines such as IL-2 and IFN-γ. This data shows the selective role of ACS in the upregulation of the Th2 response and suggests its use for succeeding selective Th2 immunomodulation (Bani et al. 2011). Some reports also propose that the cytotoxic activity of saffron extract may be related to a decrease in nitric oxide (NO) concentration produced by the hepatocellular carcinoma cell line (HepG-2) and laryngeal carcinoma cell line (Hep-2) (Parizadeh et al. 2011).

3.7 Antioxidant Activity

The antioxidant activities of *C. sativus* and its apocarotenoids are well reported in literature (Baba et al. 2015a). The ROS scavenging potential of crocin is considered responsible for its various pharmacological properties which include neuroprotective, anti-inflammatory, and anticancer activities. Bors et al. (1982) demonstrated participation of free radicals in bleaching of aqueous solution of crocin. Afterward, this property was used to determine the antioxidant activity (Bors et al. 1984). Extensively investigated, crocin bleaching inhibition assay is used to evaluate the antioxidant activity of many substances (Bathaie et al. 2011; Bors et al. 1984; Bountagkidou et al. 2012; Hosseinzadeh and Jahanian 2010; Kampa et al. 2002; Tubaro et al. 1996). Crocin isolated from *Gardenia* fruit has also been reported to possess antioxidant capacity at low concentrations (Han et al. 1994). Another study has revealed that the antioxidant activity of crocin is determined through thiocyanate method and found to be better than that of TBA (Pham et al. 2000). In this study methanolic extract of crocin isolated from *C. sativus* L. exhibited ROS scavenging (Assimopoulou et al. 2005). A GSH-dependent mechanism has been reported to participate in crocin-preventive effects on cell death induced by oxidative stress by Soeda et al. (2007). A close association between total crocin content and antioxidant activity of saffron has been reported in several in vitro studies. However, this antioxidant capacity has been reported to be heavily affected by the degree of glycosylation in the crocin structures (Chen et al. 2010b). The ethanolic and aqueous extracts of *C. sativus*, crocin, and safranal have been reported to diminish the extent of MDA production. It also revealed that the preventive effect of crocin on liver microsomal lipid peroxidation was as par with that of BHT (100 µM) (Hosseinzadeh et al. 2009b). Ordoudi et al. (2009) showed low antioxidant activity of saffron extract in cell-free systems in comparison with well-known antioxidants. However, its capacity to decrease intracellular reactive oxygen species generation was as efficient as that of phenolic antioxidants. In another study the role of crocin in the treatment of secondary problems associated with snakebite has also been reported. Crocin ameliorates pro-inflammatory cytokine levels such as L-1b, TNF-α, and IL-6 and reduces the venom-induced oxidative stress (Sebastin Santhosh et al. 2012). Crocin is also considered as one of the major contributor to the antioxidant properties of saffron extracts with the involvement of several mechanisms such as nitrite scavenging ability, radical cation suppression, superoxide dismutase-like activity, and elongation of lipid peroxidation (Yang et al. 2011). Moreover, it is believed that the antioxidant potential of saffron and its apocarotenoids is responsible for most of its pharmacological properties; however, more research endeavors are urgently required.

The antioxidant effect of crocin 1 and crocetin in Kunming mice administered orally was observed to be equally efficacious (Chen et al. 2010a). In yet another investigation, it has been observed that pretreatment rats intoxicated by $BeCl_2$ with crocin lead to significant increase in the transcript levels of CAT and SOD (El-Beshbishy et al. 2012). Besides crocin and crocetin, other apocarotenoids of *C. sativus* have also been shown to possess significant antioxidant activity. Assimopoulou et al. (2005) observed that safranal (500 ppm in methanol sol) exhibits 34 % radical scavenging. This is most probably attributed to its capacity to donate a hydrogen atom to DPPH radical. Different in vitro methods such as deoxyribose have been used to evaluate the antioxidant properties of safranal, and it has been observed that safranal exhibited considerable hydroxyl radical scavenging property in deoxyribose assay dose-dependently and also reduced the generation of MDA in RBCs, lipid peroxidation induced by hydrogen peroxide, and nonenzymatic lipid peroxidation in liver microsomes (Hosseinzadeh et al. 2009b).

3.8 Effects of *C. sativus* and Its Constituents on Anxiety

Anxiety is defined as an emotional anticipation of an aversive situation and is reflected by species-specific behavioral fear responses to threatening and stressful stimuli. Some of anxiety disorders are a major public health issue worldwide. These include generalized anxiety disorder (GAD), specific and social phobias, post-traumatic stress disorder (PTSD), obsessive–compulsive disorder (OCD), and panic disorder. Generally, anxiety disorders are treated with medications targeting γ-aminobutyric acid (GABA) and serotonergic neurotransmission including benzodiazepines, partial agonists of the serotonergic 5-HT1A receptor, and selective serotonin reuptake inhibitors (SSRIs).

Moreover, benzodiazepines or SSRIs may be accompanied with some harmful side effects. These side effects include sedation, memory deficits, dependence and withdrawal, weight gain, and sexual dysfunction (Tarantilis et al. 1995; Hammer et al. 2004; Cryan and Sweeney 2011; Gorman 2003).

The effects of *C. sativus* and its constituents on anxiety have been widely reported in literature. There are strong evidences that indicate anxiolytic-like effect of crocins in procedures assessing anxiety in rats. Administration of 50 mg/kg crocins in the light/dark test enhanced the time to enter the dark compartment. However, it did not affect the number of shifts between the light and the dark chamber and extended the time spent in the lit compartment of the light/dark box. This effect was similar to the standard molecule diazepam (1.5 mg/kg). Moreover, crocins (30 and 50 mg/kg) were reported to reduce compulsive behavior (excessive grooming) induced by the serotonergic 5-HT2c receptor agonist 1-(3-chlorophenyl)piperazine hydrochloride (mCPP) (0.6 mg/kg). It should be noted that crocins did not affect locomotor activity of rats at any tested dose. Taken together, these investigations indicate that the active components of saffron bring anxiolytic-like behavior in rats and the anxiolytic activity should not be accredited to variations in locomotor activity (Georgiadou et al. 2012; Hosseinzadeh and Noraei 2009).

In yet another investigation on mice, it has been observed that *C. sativus* aqueous extracts and its constituents, safranal and crocin, caused an anxiolytic-like effect almost similar to that of diazepam. However, at high doses, the antianxiety effects were not observed. It is noteworthy that the saffron aqueous extracts lead to sedation by diminishing mice motility and motor coordination as observed in open field and rotarod tests. Further, safranal at a low dose range also leads to hypomotility and enhanced grooming and rearing. However, crocin (50–600 mg/kg) did not change mice behavior in the elevated plus maze test and at 200 and 600 mg/kg decreased mice motility (Hosseinzadeh and Noraei 2009). The results of this latter study (Hosseinzadeh and Noraei 2009) seem to be in disparity with the above observations (Gorman 2003; Pitsikas et al. 2008a, b) wherein an anxiolytic activity of crocins was observed. These differences in the results may be attributed to different experimental setups.

The mechanisms by which *C. sativus* and its active constituents apply their anxiolytic activity similar in magnitude to that of benzodiazepine diazepam are still being investigated. Benzodiazepines exert their anxiolytic activity by indirectly acting as agonists on GABAA receptor. It has also been reported that a few plant flavonoids exhibit an affinity to benzodiazepine binding site at the GABAA receptor (Ai et al. 1997; Marder et al. 2001). It may therefore be hypothesized that saffron and its constituents may also exert their anxiolytic effect by competing for the benzodiazepine binding site of GABAA receptor. However, further investigations are needed to confirm the hypothesis. Moreover, the mechanism that is responsible for antistress activity of *C. sativus* and its constituents is yet to be elucidated completely. There are certain concrete reports that propose stress-induced activation of hypothalamus–pituitary–adrenal (HPA) axis leading to the plasma corticosterone augmentation as a response (Miller and O'Callaghan 2002). According to recent studies (Halatei et al. 2011), mice that were administrated with aqueous extracts of saffron or crocin did not demonstrate an increase of plasmatic corticosterone levels during stress. It is however possible that crocin interacts with the HPA axis and decreases the stress-induced corticosterone upsurge (Halatei et al. 2011). In connection with these reports, it has been proposed that saffron can inhibit N-methyl-D-aspartate (NMDA) and sigma opioid receptors (Lechtenberg et al. 2008). The latter is of interest as NMDA and sigma receptors can control corticosterone discharge from the adrenal cortex (Iyengar et al. 1990). It can be therefore concluded that *C. sativus* and its major apocarotenoid, crocin, may prevent corticosterone release in stressed mice through blockade of NMDA and/or sigma opioid receptors (Halatei et al. 2011).

3.9   Effects of *C. sativus* and Its Constituents in Schizophrenia

Schizophrenia is considered as one of the serious mental disorders. It affects up to 1 % of the population around the globe. It harms social, professional, and individual functioning resulting in a significant deterioration in the quality of life. The etiology and pathophysiology of schizophrenia is still not understood. Schizophrenic patients suffer from long-term and continuous psychotic problems. The symptoms can be categorized into three main types: (1) positive symptoms which include hallucinations or imaginations, delusions or false impression, disordered thought processing, and catatonic behavior; (2) negative symptoms which include social isolation, anhedonia or inability to feel pleasure, and avolition; and (3) cognitive disturbances which include attention deficits and memory (Freedman 2003). Anomalies in several neurotransmitters, mostly dopamine, glutamate, cholinergic, serotonergic, and GABAergic systems, are believed to be central for the advent of schizophrenia (Steeds et al. 2015). While on the one hand positive symptoms of this disease are accompanied by surplus dopaminergic neurotransmission in striatal brain regions, on the other hand negative symptoms and cognitive deficits are associated with reduced dopaminergic functioning.

The effects of saffron and its components on schizophrenia have also been investigated. Studies have reported that acute application of crocins (15–30 mg/kg) retreated recognition memory deficits caused by the NMDA receptor antagonist ketamine (3 mg/kg) in rats providing a clue toward the role of crocins in schizophrenia-related cognitive discrepancies. Moreover, crocins (50 mg/kg) weakened ketamine (25 mg/kg)-induced psychotomimetic effects (hypermotility, stereotypies, and ataxia) in the rat. The active constituents of *C. sativus* (50 mg/kg) were also observed to weaken the social isolation prompted by sub-chronic treatment with ketamine (8 mg/kg) in rats using social interaction test. Till date, only one clinical trial is reported in the literature that was carried out to evaluate the safety and permissibility of *C. sativus* and its apocarotenoids in schizophrenia. A double-blind placebo-controlled trail was done on 61 schizophrenia patients. The schizophrenics received treatment twice daily (saffron or crocin 15 mg) or placebo for 12 consecutive weeks. The study indicated that *C. sativus* extracts and crocin given at 15 mg twice daily were safe and tolerable in schizophrenic patients (Akhondzadeh et al. 2010a, 2010b; Farokhnia et al. 2014; Mousavi et al. 2015). The mechanism by which crocins exert their effects on ketamine-induced behavioral discrepancies is not well understood, and therefore additional studies need to be carried out to elucidate the proper mechanism. It is however well reported that schizophrenia-like effects of NMDA1 receptor antagonists (e.g., ketamine) include augmented levels of glutamate, hypermotility, stereotypy, and cognition discrepancies (Moghaddam et al. 1997). In this perspective, it has been observed that acute systemic administration of safranal lessened kainic acid-induced enhancement of extracellular glutamate concentrations in the rat hippocampus (Hosseinzadeh et al. 2008a, b). Moreover, it has been shown that saffron extracts prevent glutamatergic synaptic transmission in brain (Berger et al. 2011). Taken together, these studies suggest that this decrease of glutamate levels by *C. sativus* and its components might account for the beneficial effects exhibited by crocins on ketamine-induced behavioral discrepancies.

3.10 Conclusion

Saffron and its ingredients possess a diversity of activities including anticancer, cardioprotective, neuroprotective, and many others. Most of these activities of saffron may be attributed to the antioxidant activities of its apocarotenoid constituents. Moreover, the suppressive effects of saffron are partly due to the anti-inflammatory properties of the crocin by the inhibition of several cytokines and inducible inflammatory enzymes. Currently, evaluation of suitable intake doses of saffron apocarotenoids without cytotoxicity needs to be determined. Further, toxicological studies carried on saffron are not clear and require further endeavors.

References

Abdullaev FI, Espinosa-Aguirre JJ (2004) Biomedical properties of saffron and its potential use in cancer therapy and chemoprevention trials. Cancer Detect Prev 28:426–432

Abdullaev FI, Frenkel GD (1999) Saffron in biological and medical research. Saffron: Crocus sativus L. Harwood Academic Publishers, Amsterdam, pp 103–114

Abdullaev FI, Riveron-Negrete L, Caballero-Ortega H, Hernández JM, Perez-Lopez I, Pereda-Miranda R, Espinosa-Aguirre JJ (2003) Use of in vitro assays to assess the potential antigenotoxic and cytotoxic effects of saffron (Crocus sativus L.). Toxicol in Vitro 17(5):731–736

Abe K, Sugiura M, Yamaguchi S, Shoyama Y, Saito H (1999) Saffron extract prevents acetaldehyde-induced inhibition of long-term potentiation in the rat dentate gyrus in vivo. Brain Res 851(1):287–289

Ai J, Dekermendjian K, Wang X, Nielsen M, Witt MR (1997) 6-Methylflavone, a benzodiazepine receptor ligand with antagonistic properties on rat brain and human recombinant GABAA receptors in vitro. Drug Dev Res 41(2):99–106

Akhondzadeh S, Tahmacebi-Pour N, Noorbala A, Amini H, Fallah-Pour H, Jamshidi A et al (2005) Crocus sativus L. in the treatment of mild to moderate depression: a double blind, randomized and placebo controlled trial. Phytother Res 19:148–151

Akhondzadeh S, Sabet MS, Harirchian MH, Togha M, Cheraghmakani H, Razeghi S, … Rezazadeh SA (2010a) A 22-week, multicenter, randomized, double-blind controlled trial of Crocus sativus in the treatment of mild-to-moderate Alzheimer's disease. Psychopharmacology 207(4):637–643

Akhondzadeh S, Sabet MS, Harirchian MH, Togha M, Cheraghmakani H, Razeghi S, … Zare F (2010b) Saffron in the treatment of patients with mild to moderate Alzheimer's disease: a 16-week, randomized and placebo-controlled trial. J clin phar ther 35(5):581–588

Alavizadeh SH, Hosseinzadeh H (2014) Bioactivity assessment and toxicity of crocin: a comprehensive review. Food Chem Toxicol 64:65–80

Al-Qudsi F, Ayedh A (2012) Effect of saffron on mouse embryo development. J Am Sci 8:1554–1568

Altinoz E, Oner Z, Elbe H, Cigremis Y, Turkoz Y (2015) Protective effects of saffron (its active constituent, crocin) on nephropathy in streptozotocin-induced diabetic rats. Hum Exp Toxicol 34(2):127–134

Assimopoulou AN, Sinakos Z, Papageorgiou VP (2005) Radical scavenging activity of Crocus sativus L. extract and its bioactive constituents. Phytother Res 19:997–1000

Aung HH, Wang CZ, Ni M, Fishbein A, Mehendale SR, Xie JT, Shoyama CY, Yuan CS (2007) Crocin from *Crocus sativus* possesses significant antiproliferation effects on human colorectal cancer cells. Exp Oncol 29:175–180

Baba SA, Malik AH, Wani ZA, Mohiuddin T, Shah Z, Abbas N, Ashraf N (2015) Phytochemical analysis and antioxidant activity of different tissue types of Crocus sativus and oxidative stress alleviating potential of saffron extract in plants, bacteria, and yeast. S Afr J Bot 99:80–87

Bahashwan S, Hassan MH, Aly H, Ghobara MM, El-Beshbishy HA, Busati I (2015) Crocin mitigates carbon tetrachloride-induced liver toxicity in rats. J Taibah Univ Med Sci 10(2):140–149

Bakshi H, Sam S, Rozati R, Sultan P, Islam T, Rathore B, … Saxena RC (2010) DNA fragmentation and cell cycle arrest: a hallmark of apoptosis induced by crocin from kashmiri saffron in a human pancreatic cancer cell line. Asian Pac J Cancer Prev 11(3):675–679

Bani S, Pandey A, Agnihotri VK, Pathania V, Singh B (2011) Selective Th2 upregulation by Crocus sativus: a neutraceutical spice. Evid Based Complement Alternat Med 2011:1–9

Bathaie SZ, Kermani FMZ, Shams A (2011) Crocin bleaching assay using purified di- gentiobiosyl crocin (a-crocin) from Iranian saffron. Iran J Basic Med Sci 14:399–406

Bathaie SZ, Miri H, Mohagheghi MA, Mokhtari-Dizaji M, Shahbazfar AA, Hasanzadeh H (2013) Saffron aqueous extract inhibits the chemically-induced gastric cancer progression in the wistar albino rat. Iran J basic Med Sci 16(1):26–38

Berger F, Hensel A, Nieber K (2011) Saffron extracts and trans-crocetin inhibit glutamatergic synaptic transmission in rat cortical brain slices. Neuroscience 180:238–247

Bhargava V (2011) Medicinal uses and pharmacological properties of *Crocus sativus* Linn (Saffron). Int J Pharm Pharm Sci 3(3):22–26

Bors W, Saran M, Michel C (1982) Radical intermediates involved in the bleaching of the carotenoid crocin. Hydroxyl radicals, superoxide anions and hydrated electrons. Int J Radiat Biol Relat Stud Phys Chem Med 41:493–501

Bors W, Michel C, Saran M (1984) Inhibition of the bleaching of the carotenoid crocin a rapid test for quantifying antioxidant activity. Biochim Biophys Acta 796:312–319

Bountagkidou O, Van der Klift EJC, Tsimidou MZ, Ordoudi SA, Van Beek TA (2012) An on-line high performance liquid chromatography-crocin bleaching assay for detection of antioxidants. J Chromatogr A 1237:80–85

Chen Y, Yang T, Huang J, Tian X, Zhao C, Cai L, Feng LJ, Zhang H (2010a) Comparative evaluation of the antioxidant capacity of crocetin and crocin in vivo. Chin Pharmacol Bull 26:248–251

Chen Y, Zhang H, Li YX, Cai L, Huang J, Zhao C, Jia L, Buchanan R, Yang T, Jiang LJ (2010b) Crocin and geniposide profiles and radical scavenging activity of gardenia fruits (*Gardenia jasminoides* Ellis) from different cultivars and at the various stages of maturation. Fitoterapia 81:269–273

Chen B, Hou ZH, Dong Z, Li CD (2015) Crocetin downregulates the proinflammatory cytokines in methylcholanthrene-induced rodent tumor model and inhibits COX-2 expression in cervical cancer cells. BioMed Res Int 2015:1–5

Chew BP, Park JS (2004) Carotenoid action on the immune response. J Nutr 134(1):257S–261S

Chryssanthi DG, Lamari FN, Iatrou G, Pylara A, Karamanos NK, Cordopatis P (2007) Inhibition of breast cancer cell proliferation by style constituents of different Crocus species. Anticancer Res 27:357–362

Cryan JF, Sweeney FF (2011) The age of anxiety: role of animal models of anxiolytic action in drug discovery. Br J Pharmacol 164:1129–1161

Dianat M, Esmaeilizadeh M, Badavi M, Samarbaf-zadeh AR, Naghizadeh B (2014) Protective effects of crocin on ischemia-reperfusion induced oxidative stress in comparison with vitamin E in isolated rat hearts. Jundishapur J Nat Pharm Prod 9(2):e17187

Dong X, Song YN, Liu WG, Guo XL (2009) Mmp-9, a potential target for cerebral ischemic treatment. Curr Neuropharmacol 7:269

Du P, Qian Z, Shen X, Rao S, Wen N (2005) Effectiveness of crocin against myocardial injury. Chin New Drugs J 14:1424

El-Beshbishy HA, Hassan MH, Aly HAA, Doghish AS, Alghaithy AAA (2012) Crocin "saffron" protects against beryllium chloride toxicity in rats through diminution of oxidative stress and enhancing gene expression of antioxidant enzymes. Ecotoxicol Environ Saf 83:47–54

EL-Maraghy SA, Rizk SM, El-Sawalhi MM (2009) Hepatoprotective potential of crocin and curcumin against iron overload-induced biochemical alterations in rat. Afr J Biochem Res 3:215–221

Escribano J, Alonso GL, Coca-Prados M, Fernandez JA (1996) Crocin, safranal and picrocrocin from saffron (Crocus sativus L.) inhibit the growth of human cancer cells in vitro. Cancer Lett 100:23–30

Farnsworth NR (1994) Ethnobotany and the search for new drugs. Wiley, Chichester

Farokhnia M, Shafiee Sabet M, Iranpour N, Gougol A, Yekehtaz H, Alimardani R, … Akhondzadeh S (2014) Comparing the efficacy and safety of Crocus sativus L. with memantine in patients with moderate to severe Alzheimer's disease: a double-blind randomized clinical trial. Hum Psychopharmacol Clin Exp 29(4):351–359

Freedman R (2003) Schizophrenia. N Engl J Med 349:1738–1749

Garc-Olmo DC, Riese HH, Escribano J, Onta J, Fernandez JA, Atiénzar M, Garcí-Olmo D (1999) Effects of long-term treatment of colon adenocarcinoma with crocin, a carotenoid from saffron (*Crocus sativus* L.): an experimental study in the rat. Nutr Cancer 35:120–126

Georgiadou G, Tarantilis PA, Pitsikas N (2012) Effects of the active constituents of Crocus sativus L.; crocins in an animal model of obsessive-compulsive disorder. Neurosci Lett 528:27–30

Geromichalos GD, Lamari FN, Papandreou MA, Trafalis DT, Margarity M, Papageorgiou A et al (2012) Saffron as a source of novel acetylcholinesterase inhibitors: molecular docking and in vitro enzymatic studies. J Agric Food Chem 60:6131–6138

Gorman JM (2003) New molecule targets for antianxiety interventions. J Clin Psychiatry 64:28–35

Goyal SN, Arora S, Sharma AK, Joshi S, Ray R, Bhatia J, … Arya DS (2010) Preventive effect of crocin of Crocus sativus on hemodynamic, biochemical, histopathological and ultrastructural alterations in isoproterenol-induced cardiotoxicity in rats. Phytomedicine 17(3):227–232

Gresta F, Avola G, Lombardo GM, Siracusa L, Ruberto G (2009) Analysis of flowering, stigmas yield and qualitative traits of saffron (Crocus sativus L.) as affected by environmental conditions. Sci Hortic 119(3):320–324

Halatei BS, Khosravi M, Sahrei H, Golmanesch L, Zardooz H, Jalili C, Ghoshoomi H (2011) Saffron (*Crocus sativus*) aqueous extract and its constituent crocin reduces stress-induced anorexia in mice. Phytother Res 25:1833–1838

Hammer MB, Robert S, Fruech BS (2004) Treatment-resistant posttraumatic stress disorder: strategies for intervention. CNS Spectr 9:740–752

Han YN, Oh HK, Hwang KH, Lee MS (1994) Antioxidant components of Gardenia fruit. Kor J Pharm 25:226–232

Hassani FV, Naseri V, Razavi BM, Mehri S, Abnous K, Hosseinzadeh H (2014) Antidepressant effects of crocin and its effects on transcript and protein levels of CREB, BDNF, and VGF in rat hippocampus. DARU J Pharm Sci 22(1):1

He SY, Qian ZY, Tang FT, Wen N, Xu GL, Sheng L (2005) Effect of crocin on experimental atherosclerosis in quails and its mechanisms. Life Sci 77(8):907–921

He SY, Qian ZY, Wen N, Tang FT, Xu GL, Zhou CH (2007a) Influence of crocetin on experimental atherosclerosis in hyperlipidemic-diet quails. Eur J Pharmacol 554(2–3):191–195

He SY, Qian ZY, Wen N, Tang FT, Xu GL, Zhou CH (2007b) Influence of crocetin on experimental atherosclerosis in hyperlipidemic-diet quails. Eur J Pharmacol 554:191–195

Himeno H, Sano K (1987) Synthesis of crocin, picrocrocin and safranal by saffron stigma-like structures proliferated in vitro. Agric Biol Chem 9(51):2395–2400

Hosseinzadeh H, Jahanian Z (2010) Effect of Crocus sativus L. (saffron) stigma and its constituents, crocin and safranal, on morphine withdrawal syndrome in mice. Phytother Res 24(5):726–730

Hosseinzadeh H, Nassiri-Asl M (2013) Avicenna's (Ibn Sina) the canon of medicine and saffron (*Crocus sativus*): a review. Phytother Res 27(4):475–483

Hosseinzadeh H, Noraei NB (2009) Anxiolytic and hypnotic effect of *Crocus sativus* aqueous extract and its constituent, crocins and safranal in mice. Phytother Res 23:768–774

Hosseinzadeh H, Ziaei T (2006) Effects of *Crocus sativus* stigma extract and its constituents, crocin and safranal, on intact memory and scopolamine-induced learning deficits in rats performing the Morris water maze task. J Med Plants 3(19):40–50

Hosseinzadeh H, Motamedshariaty V, Hadizadeh F (2007) Antidepressant effect of kaempferol, a constituent of saffron (*Crocus sativus*) petal, in mice and rats. Pharmacologyonline 2:367–370

Hosseinzadeh H, Sadeghnia HR, Rahimi A (2008a) Effects of safranal on extracellular hippocampal levels of glutamate and aspartate during kainic acid treatment in anesthetized rats. Planta Med 74:1441–1445

Hosseinzadeh H, Ziaee T, Sadeghi A (2008b) The effect of saffron, Crocus sativus stigma, extract and its constituents, safranal and crocin on sexual behaviors in normal male rats. Phytomedicine 15:491–495

Hosseinzadeh H, Modaghegh MH, Saffari Z (2009a) *Crocus sativus* L. (Saffron) extract and its active constituents (crocin and safranal) on ischemia-reperfusion in rat skeletal muscle. Evid Based Complement Alternat Med 6:343–350

Hosseinzadeh H, Shamsaie F, Mehri S, (2009b) Antioxidant activity of aqueous and ethanolic extracts of *Crocus sativus* L. stigma and its bioactive constituents, crocin and safranal. Pharmacogn Mag 5:419

Hosseinzadeh H, Sadeghnia HR, Ghaeni FA, Motamedshariaty VS, Mohajeri SA (2012) Effects of saffron (*Crocus sativus* L.) and its active constituent, crocin, on recognition and spatial memory after chronic cerebral hypoperfusion in rats. Phytother Res 26:381–386

Imenshahidi M, Zafari H, Hosseinzadeh H (2011) Effects of crocin on the acquisition and reinstatement of morphine-induced conditioned place preference in mice. Pharmacologyonline 1:1007–1013

Inoue E, Shimizu Y, Shoji M, Tsuchida H, Sano Y, Ito C (2005) Pharmacological properties of N-095, a drug containing red ginseng, polygala root, saffron, antelope horn and aloe wood. Am J Chin Med 33(1):49–60

Iyengar S, Mick S, Dilworth V, Michel J, Rao TS, Farah JM, Wood PL (1990) Sigma receptors modulate the hypothalamic-pituitary-adrenal (HPA) axis centrally: evidence for a functional interaction with NMDA receptors, in vivo. Neuropharmacology 29(3):299–303

Jain P, Pareek A, Ratan Y, Sharma S, Paliwal S (2013) Free radicals and dietary antioxidants: a potential review. Int J Pharm Sci Rev Res 18:34–48

Kadian SS, Garg M (2012) Pharmacological effects of carotenoids: a review. Int J Pharm Sci Res 3(1):42

Kampa M, Nistikaki A, Tsaousis V, Maliaraki N, Notas G, Castanas E (2002) A new automated method for the determination of the Total Antioxidant Capacity (TAC) of human plasma, based on the crocin bleaching assay. BMC Clin Pathol 2:1–16

Karimi G, Taiebi N, Hosseinzadeh H, Shirzad F (2004) Evaluation of subacute toxicity of aqueous extract of *Crocus sativus* L. stigma and petal in rats. J Med Plants 3:29–35

Kawabata K, Tung NH, Shoyama Y, Sugie S, Mori T, Tanaka T (2012) Dietary crocin inhibits colitis and colitis-associated colorectal carcinogenesis in male ICR mice. Evid-Based Complement Alternat Med 2012:1–3

Kianbakht S, Mozaffari K (2009) Effects of saffron and its active constituents, crocin and safranal, on prevention of indomethacin induced gastric ulcers in diabetic and nondiabetic rats. ‫ﻓﺼﻠﻨﺎﻣﻪ ﻋﻠﻤﻰ ﭘﮋوﻫﺸﻰ ﮔﯿﺎﻫﺎن داروﯾﻰ‬, 1(29):30–38

Konoshima T, Takasaki M, Tokuda H, Morimoto S, Tanaka H, Kawata E, Xuan L, Saito H, Sugiura M, Molnar J (1998) Crocin and crocetin derivatives inhibit skin tumour promotion in mice. Phytother Res 12:400–404

Lechtenberg M, Schepmann D, Niehues M, Hellenbrand N, Wunsch B, Hensel A (2008) Quality and functionality of saffron: quality control, species assortment and affinity of extract and isolated saffron compounds to NMDA and sigma~ 1 (sigma-1) receptors. Planta Med 74(7):764

Lee IA, Lee JH, Baek NI, Kim DH (2005) Antihyperlipidemic effect of crocin isolated from the fructus of Gardenia jasminoides and its metabolite crocetin. Biol Pharm Bull 28:2106–2110

Liu J, Qian Z (2005) Effects of crocin on cholestane-3beta-5alpha-6beta-triol-induced apoptosis and related gene expression of cultured endothelial cells. J Chin Pharm Univ 36:254

Lv CF, Luo CL, Ji HY, Zhao P (2008) Influence of crocin on gene expression profile of human bladder cancer cell lines T24. Zhongguo Zhong yao za zhi= Zhongguo zhongyao zazhi= China journal of Chinese materia medica 33(13):1612–1617

Marder M, Estiú G, Blanch LB, Viola H, Wasowski C, Medina JH, Paladini AC (2001) Molecular modeling and QSAR analysis of the interaction of flavone derivatives with the benzodiazepine binding site of the GABA A receptor complex. Bioorg Med Chem 9(2):323–335

Mehri S, Abnous K, Mousavi SH, Shariaty VM, Hosseinzadeh H (2012) Neuroprotective effect of crocin on acrylamide-induced cytotoxicity in PC12 cells. Cell Mol Neurobiol 32:227–235

Miller DB, O'Callaghan JP (2002) Neuroendocrine aspects of the response to stress. Metabolism 51(6):5–10

Moghaddam B, Adams B, Verma A, Daly D (1997) Activation of glutamatergic neurotransmission by ketamine: a novel step in the pathway from NMDA receptor blockade to dopaminergic and cognitive disruptions associated with the prefrontal cortex. J Neurosci 17(8):2921–2927

Moghaddasi MS (2010) Saffron chemicals and medicine usage. J Med Plants 4(6):427–430

Mousavi SH, Tayarani NZ, Parsaee H (2010) Protective effect of saffron extract and crocin on reactive oxygen species-mediated high glucose-induced toxicity in pc12 cells. Cell Mol Neurobiol 30:185–191

Mousavi SH, Moallem SA, Mehri S, Shahsavand S, Nassirli H, Malaekeh-Nikouei B (2011) Improvement of cytotoxic and apoptogenic properties of crocin in cancer cell lines by its nano-liposomal form. Pharm Biol 49(10):1039–1045

Mousavi B, Bathaie SZ, Fadai F, Ashtari Z (2015) Safety evaluation of saffron stigma (Crocus sativus L.) aqueous extract and crocin in patients with schizophrenia. Avicenna J Phytomedicine 5(5):413

Namin MH, Ebrahimzadeh H, Ghareyazie B, Radjabian T, Gharavi S, Tafreshi N (2009) In vitro expression of apocarotenoid genes in *Crocus sativus* L. Afr J Biotechnol 8(20):5378–5382

Noureini SK, Wink M (2012) Antiproliferative effects of crocin in HepG2 cells by telomerase inhibition and hTERT down-regulation. Asian Pac J Cancer Prev 13(5):2305–2309

Ochiai T, Soeda S, Ohno S, Tanaka H, Shoyama Y, Shimeno H (2004a) Crocin prevent the death of PC-12 cells through sphingomyelinase-ceramide signaling by increasing glutathione synthesis. Neurochem Int 44:321–330

Ochiai T, Ohno S, Soeda S, Tanaka H, Shoyama Y, Shimeno H (2004b) Crocin prevents the death of rat pheochromocytoma (PC-12) cells by its antioxidant effects stronger than those of a-tocopherol. Neurosci Lett 362:61–64

Ordoudi SA, Befani CD, Nenadis N, Koliakos GG, Tsimidou MZ (2009) Further examination of antiradical properties of crocus sativus stigmas extract rich in crocins. J Agric Food Chem 57:3080–3086

Papandreou MA, Tsachaki M, Efthimiopoulos S, Cordopatis P, Lamari FN, Margarity M (2011) Memory enhancing effects of saffron in aged mice are correlated with antioxidant protection. Behav Brain Res 219(2):197–204

Parizadeh MR, Ghafoori Gharib F, Abbaspour AR, Tavakol Afshar J, Ghayour-Mobarhan M (2011) Effects of aqueous saffron extract on nitric oxide production by two human carcinoma cell lines: hepatocellular carcinoma (HepG2) and laryngeal carcinoma (Hep2). Avicenna J Phytomedicine 1(1):43–50

Perez-Rodriguez L, Mougeot F, Alonso-Alvarez C, Blas J, Viñuela J, Bortolotti GR (2008) Cell-mediated immune activation rapidly decreases plasma carotenoids but does not affect oxidative stress in red-legged partridges (*Alectoris rufa*). J Exp Biol 211(13):2155–2161

Pfister S, Meyer P, Steck A, Pfander H (1996) Isolation and structure elucidation of carotenoid-glycosyl esters in gardenia fruits (Gardenia jasminoides Ellis) and saffron (*Crocus sativus* Linne). J Agric Food Chem 44(9):2612–2615

Pham TQ, Cormier F, Farnworth E, Tong VH, Van Calsteren MR (2000) Antioxidant properties of crocin from Gardenia jasminoides Ellis and study of the reactions of crocin with linoleic acid and crocin with oxygen. J Agric Food Chem 48:1455–1461

Pitsikas N, Zisopoulou S, Tarantilis PA, Kanakis CD, Polissiou MG, Sakellaridis N (2007) Effects of the active constituents of Crocus sativus L., crocins on recognition and spatial rats memory. Behav Brain Res 183:141–146

Pitsikas N, Boultadakis A, Georgiadou G, Tarantilis PA, Sakellaridis N (2008) Effects of the active constituents of *Crocus sativus* L., crocins, in an animal model of anxiety. Phytomedicine 15:1135–1139

Pratt J, Winchester C, Dawson N, Morris B (2012) Advancing schizophrenia drug discovery: optimizing rodent models to bridge the translational gap. Nat Rev Drug Discov 11(7):560–579

Raina BL, Agarwal SG, Bhatia AK, Gaur GS (1996) Changes in pigments and volatiles of saffron (*Crocus sativus* L.) during processing and storage. J Sci Food Agric 71:27–32

Rastgoo M, Hosseinzadeh H, Alavizadeh H, Abbasi A, Ayati Z, Jaafari MR (2013) Antitumor activity of PEGylated nanoliposomes containing crocin in mice bearing C26 colon carcinoma. Planta Med 79:447–451

Rezaee R, Hosseinzadeh H (2013) Safranal: from an aromatic natural product to a rewarding pharmacological agent. Iran J Basic Med Sci 16(1):12

Richelson E (1994) Pharmacology of antidepressants characteristics of the ideal drug. Mayo Clin Proc 69:1069–1081

Sadeghnia HR, Cortez MA, Liu D, Hosseinzadeh H, Carter Snead O (2008) Antiabsence effects of safranal in acute experimental seizure models: EEG and autoradiography. J Pharm Pharm Sci 11:1–14

Saleem S, Ahmad M, Ahmad AS, Yousuf S, Ansari MA, Khan MB, Ishrat T, Islam F (2006) Effect of saffron (Crocus sativus) on neurobehavioral and neurochemical changes in cerebral ischemia in rats. J Med Food 9:246–253

Shen XC, Qian ZY, Chen Q, Wang YJ (2004) Protective effect of crocetin on primary culture of cardiac myocyte treated with noradrenaline in vitro. Acta Pharm Sin 39(10):787–791

Sheng L, Qian Z, Zheng S, Xi L (2006) Mechanism of hypolipidemic effect of crocin in rats: crocin inhibits pancreatic lipase. Eur J Pharmacol 543:116–122

Shiping M, Baolin L, Sudi Z (1999) Pharmacological studies of glycosides of saffron Crocus (Crocus sativus) effects on blood coagulation, platelet aggregation and thrombosis. Chin Trad Herb Drugs 3:196–198

Soeda S, Ochiai T, Shimeno H, Saito H, Abe K, Tanaka H, Shoyama Y (2007) Pharmacological activities of crocin in saffron. J Nat Med 61:102–111

Steeds H, Carhart-Harris RL, Stone JM (2015) Drug models of schizophrenia. Ther Adv Psychopharmacol 5:43–58

Sun J, Xu XM, Ni CZ, Zhang H, Li XY, Zhang CL, Liu YR, Li SF, Zhou QZ, Zhou HM (2011) Crocin inhibits proliferation and nucleic acid synthesis and induces apoptosis in the human tongue squamous cell carcinoma cell line tca8113. Asian Pac J Cancer Prev 12:2679–2683

Talaei A, Moghadam MH, Tabassi SAS, Mohajeri SA (2015) Crocin, the main active saffron constituent, as an adjunctive treatment in major depressive disorder: a randomized, double-blind, placebo-controlled, pilot clinical trial. J Affect Disord 174:51–56

Tamaddonfard E, Gooshchi NH, Seiednejad-Yamchi S (2012) Central effect of crocin on penicillin-induced epileptiform activity in rats. Pharmacol Rep 64:94–101

Tanaka T, Shnimizu M, Moriwaki H (2012) Cancer chemoprevention by carotenoids. Molecules 17(3):3202–3242

Tarantilis PA, Tsoupras G, Polissiou M (1995) Determination of saffron (Crocus sativus L.) components in crude plant extract using high-performance liquid chromatography-UV/Visible photodiode-array detection-mass spectrometry. J Chromatogr 699:107–118

Tubaro F, Micossi E, Ursini F (1996) The antioxidant capacity of complex mixtures by kinetic analysis of crocin bleaching inhibition. J Am Oil Chem Soc 73:173–179

Wang Y, Han T, Zhu Y, Zheng CJ, Ming QL, Rahman K, Qin LP (2010) Antidepressant properties of bioactive fractions from the extract of Crocus sativus L. J Nat Med 64(1):24–30

Wang Y, Sun J, Liu C, Fang C (2014) Protective effects of crocetin pretreatment on myocardial injury in an ischemia/reperfusion rat model. Eur J Pharmacol 741:290–296

Xu G, Gong Z, Yu W, Gao L, He S, Qian Z (2007) Increased expression ratio of bcl-2/bax is associated with crocin-mediated apoptosis in bovine aortic endothelial cells. Basic Clin Pharmacol Toxicol 100:31–35

Xu GL, Li G, Ma HP, Zhong H, Liu F, Ao GZ (2009) Preventive effect of crocin in inflamed animals and in LPS-challenged RAW 264.7 cells. J Agric Food Chem 57(18):8325–8330

Xu HJ, Zhong R, Zhao YX, Li XR, Lu Y, Song AQ, … Sun LR (2010) [Proliferative inhibition and apoptotic induction effects of crocin on human leukemia HL-60 cells and their mechanisms]. Zhongguo shi yan xue ye xue za zhi/Zhongguo bing li sheng li xue hui= J Exp Hematol/Chin Assoc Pathophysiol 18(4):887–892

Xuan B, Zhou YH, Li NA, Min ZD, Chiou GC (1999) Effects of crocin analogs on ocular blood flow and retinal function. J Ocul Pharmacol Ther 15(2):143–152

Yang HJ, Park M, Lee HS (2011) Antioxidative activities and components of Gardenia jasminoides. Kor J Food Sci Technol 43:51–57

Zhao P, Luo CL, Wu XH, Hu HB, Lv CF, Ji HY (2008) Proliferation apoptotic influence of crocin on human bladder cancer T24 cell line. Zhongguo Zhong yao za zhi= Zhongguo zhongyao zazhi= China journal of Chinese materia medica 33(15):1869–1873

Zheng S, Qian Z, Tang F, Sheng L (2005) Suppression of vascular cell adhesion molecule-1 expression by crocetin contributes to attenuation of atherosclerosis in hypercholesterolemic rabbits. Biochem Pharmacol 70(8):1192–1199

Zheng YQ, Liu JX, Wang JN, Xu L (2007) Effects of crocin on reperfusion induced oxidative/nitrative injury to cerebral microvessels after global cerebral ischemia. Brain Res 1138:86–94